Modeling Ships and Space Craft

Gina Hagler

Modeling Ships and Space Craft

The Science and Art of Mastering the Oceans and Sky

 Springer

Gina Hagler
Rockville, MD
USA

ISBN 978-1-4614-4595-1   ISBN 978-1-4614-4596-8 (eBook)
DOI 10.1007/978-1-4614-4596-8
Springer New York Heidelberg Dordrecht London

Library of Congress Control Number: 2012945938

Images are illustrated by Jason Hagler unless otherwise noted.

Printed on acid-free paper

Springer is part of Springer Science+Business Media (www.springer.com)

To Grandpa Phil

Acknowledgment

My fascination with scale model testing and the science behind it began several years ago with a tour of the David Taylor Model Basin at Carderock, Maryland. During the tour our guide, Tom Warring, told my oldest son and me about the pioneering work done by Rear Admiral David Watson Taylor at the Experimental Model Basin. We'd never heard of this extraordinary man and immediately wanted to know more.

That visit to Carderock was the start of an investigation of the science and practice of model testing that began with the help of Barbara Breedan at the Nimitz Library at the United States Naval Academy. Along the way to the completion of this book, I was privileged to speak with Larrie Ferreiro, naval architect and historian, who was kind enough to arrange for Julian Simon Calero, formerly of INTA and author of "The Genesis of Fluid Mechanics," to read an early draft of what now are the first chapters.

One particularly memorable day, I met with John D. Anderson, Curator of Aerodynamics at the National Air & Space Museum at the Smithsonian Institution in Washington, DC, to discuss the wind tunnel tests conducted by Orville and Wilbur Wright. Afterward, he and I went to examine the model of the wind tunnel on display at Air & Space. I also had the opportunity to discuss the history of ship design with Dr. Horst Nowacki of the Max Planck Institute for the History of Science, model testing with William G. Day, former head of the Carderock towing basin, swim bladders with Professor of Biology Dr. Frank E. Fish, and his experience with resistance with Olympic cyclist, John Howard.

The entire process of researching was an amazing experience punctuated by the generosity of these authorities. I would like to thank them for so generously sharing their time and knowledge with me.

I was equally fortunate to have the support of those who love me, especially my three children: Jason, who drew the images for this book; Seth, who helped with all the footnotes – not once but twice; and Tess, who cheered me on when the task felt overwhelming. My thanks and love to them all!

Rockville, MD, USA   Gina Hagler

Preface

Since earliest recorded history, man has sought an understanding of natural phenomenon. Some sought understanding as a way to control phenomenon like flooding in areas vital to agriculture. Others sought understanding as a way to join in the phenomenon that intrigued them. For both those seeking to control the movement of water and those seeking to move through the air with the birds, the close observation of fluids in motion was a natural place to begin.

By the time of Aristotle, theorists had begun to record their investigations. As the years passed, observers also began to experiment with the effects of these properties on objects moving through the fluids. Each early theorist pursued the aspect of fluids that interested him, generally unaware of the work of others. Although these investigations were performed without the tools that are available today, the conclusions formed by many have proven over the centuries to be correct. By the sixteenth century, most investigations were conducted by enthusiastic amateurs or those who were to become the first in what are now established fields such as engineering. It was at this time that theories proven by replicable investigation began to take the place of long-held but unproven beliefs.

But observations and theories were one thing. The systematic application of these theories to matters of national importance was another. Even in a discipline so vital to the interest of emerging nations, it wasn't until the late nineteenth century that shipbuilders first looked to apply science to the design of ships. Prior to this, shipbuilding had been more art than science with ships built according to what had worked before and what "should" work going forward. As navies and industries exerted greater control over the design and construction process of increasingly complex and expensive vessels, there was an increased demand to demonstrate that the completed ships would satisfy a specific level of performance (e.g., speed) before governments and private owners were willing to invest the enormous money and resources needed to build these ships.

It was this dual focus on the application of science and the demand for accurate estimates of future performance that led to an examination of scale model testing as a viable method for achieving these objectives. The time was right for English engineer William Froude to champion and prove that the testing of designs on scale models in a controlled environment as a precursor to construction would yield results that were superior to those that could be achieved through a reliance on historical precedence having scant relationship to the new ships being called for. Since this time it has been accepted practice to use scale model testing to perfect the designs for new vessels before construction begins.

Scale models are used today for more than the design of ocean-going craft. They are also used for the design of aircraft and spacecraft. In areas where exact scale models are not used, prototypes often are because the value of small-scale tests in the design phase is no longer questioned. Even the sophisticated computational fluid dynamic models used to generate many vessel components in production today are based upon physical scale model testing that was completed in model basins.

All of these models predict future performance by way of the application of the fluid dynamic principles that will be in effect around the full-sized versions of these craft. This is because water, air, and gasses are all considered to be fluids. When you float a boat, fly a kite, or launch a rocket, you are putting the principles of fluid dynamics to work. Today as fluid dynamic principles are applied to problems encountered in the design of ocean vessels for best performance, to decisions about the optimal configuration of an airplane wing, and to considerations about the most efficient design for rockets and launch vehicles, the process represents the effective melding of science and innovation. This combination has now facilitated the economical and reasoned design of scores of vessels for more than 100 years.

As an appreciation for the economies linked to the use of scale models grew, the scope of their use increased greatly. An exploration of the application of scale models to the design of ocean-going vessels, aircraft, and spacecraft—along with a look at the scientific principles in action in nature and the testing facilities—will be found in the chapters that follow.

Rockville, MD, USA   Gina Hagler

Contents

Part I
Fluid Dynamics in Action

Fluid dynamic principles are in effect all around us. From the sky to the seas, they influence the way creatures and objects function. Before man could successfully fly with the birds or swim with aquatic creatures, he had to understand the mechanisms at work. To do this, he studied the animals around him.

Chapter 1
Airborne Creatures

For once you have tasted flight you will walk the earth with your eyes turned skywards, for there you have been and there you will long to return.

Leonardo da Vinci

Standing outside on a clear day with a persistent breeze, it's natural to look skyward and watch the topmost tree branches sway. Perhaps you'll observe Canada geese flying overhead in v-formation, bees buzzing past as they do their work, or dragonflies darting through the flowerbeds. You might even see Red-tailed hawks hovering far above as they ride the thermals. Whatever the season and whichever you observe, these are all instances of *fluid dynamics* at work. The same principles that apply to objects and animals as they move through the water, apply to these branches and animals as they move through the air.

Fluid dynamics is the branch of science related to the behavior of fluids in motion. *Fluids* are a state of matter in which a substance cannot maintain a shape on its own. Because of this, a fluid completely fills the space it occupies and will take the shape of the container that holds it, filling each nook and cranny. A fluid can be air, a liquid, or a gas. In the case of a fluid such as water, if the fluid does not fill the container from bottom to top, there will be an *observable surface* at the highest level (Fig. 1.1). *Dynamics* is the study of causes of motion and changes in that motion. The study of *fluid dynamics* is the study of fluids in motion. It is called *hydrodynamics* when the fluid is water, *aerodynamics* when the fluid is a gas or air.

In 1644, Evangelista Torricelli wrote, "We live submerged at the bottom of an ocean of air." Because we are surrounded by air, just as aquatic creatures are surrounded by water, the force of gravity exerts pressure on all sides. As a result of this uniform pressure, neither land nor aquatic creatures feel the full force of the pressure of the fluid around us, yet fluid dynamic principles are constantly in action. For aquatic creatures, the fluid is water and the field of study is known as *hydrodynamics*. For terrestrial creatures, the fluid is air and the field of study is known as *aerodynamics*.

G. Hagler, *Modeling Ships and Space Craft: The Science and Art of Mastering the Oceans and Sky*, DOI 10.1007/978-1-4614-4596-8_1,
© Springer Science+Business Media, LLC 2013

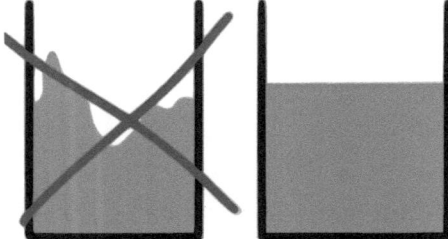

Fig. 1.2 Canada geese in v-formation. Canada geese fly in a v-formation, or skein

All birds, bats, and flying insects use fluid dynamic principles to their advantage as they make their way through the air. The Canada goose, due to its large size and wide range, is among the most obvious examples of *aerodynamics* in action when flying in its v-formation. Perhaps you've watched as a gaggle of geese floating on a pond moved to face into the wind, then ran across the water, flapping violently as they brought themselves aloft with a raucous honking. Once airborne, they formed a wide v-formation flapping in near unison as they began their long journey (Fig. 1.2).

Birds

Flight seems simple enough once you see the geese in motion overhead, but these large waterfowl have to accomplish the same essential tasks as every other airborne creature or machine as they successfully balance the four forces of flight. First, they must overcome *gravity* (the attraction of the Earth) and generate sufficient *lift* (upward motion) to counteract that gravity. With this balance attained, they must then overcame *drag* (the force of friction also known as *resistance*) and generate *thrust* (forward motion). Only when all four forces—gravity, lift, drag, and thrust—are in balance can the geese, or any object, sustain flight (Fig. 1.3).

Each step each goose took was part of this process. Even taking off into the wind was a necessary component of lift. As each goose headed into the wind, its body was tilted slightly upward to encounter the wind at a greater angle than it would if its body were more horizontal to the ground. "An airborne bird generates more lift by flying with its wings and body slightly upraised. This is called increasing the angle of attack. The air hitting the underside of the wings and belly generate an uplifting force in addition to the lift provided by Bernoulli's principle."[1]

The *angle of attack* is the angle at which the *relative wind* meets the airfoil (Fig. 1.4). It varies with the amount of lift: The larger the angle of attack, the greater the lift. There is an upper limit to the angle of attack. Known as the *stall point*, it occurs at about 17 degrees and is literally the point at which the object in flight is no longer able to proceed[2] (Fig. 1.5).

"Relative wind is created by movement of an airfoil through the air. As an example, consider a person sitting in an automobile on a no-wind day with a hand

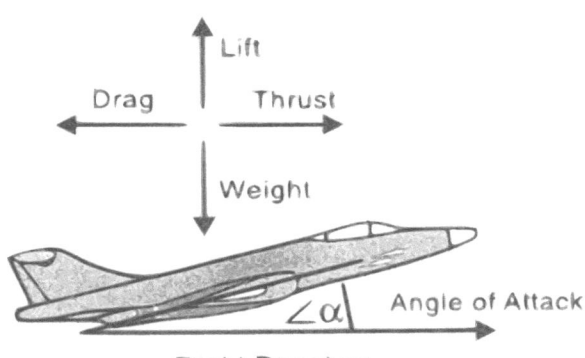

Fig. 1.3 The four forces of flight must be in balance

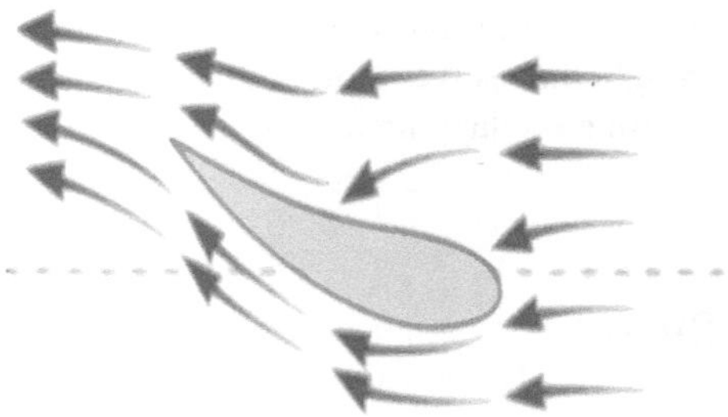

Fig. 1.4 The angle of attack is the angle at which the relative wind meets the airfoil. The greater the angle, the greater the lift

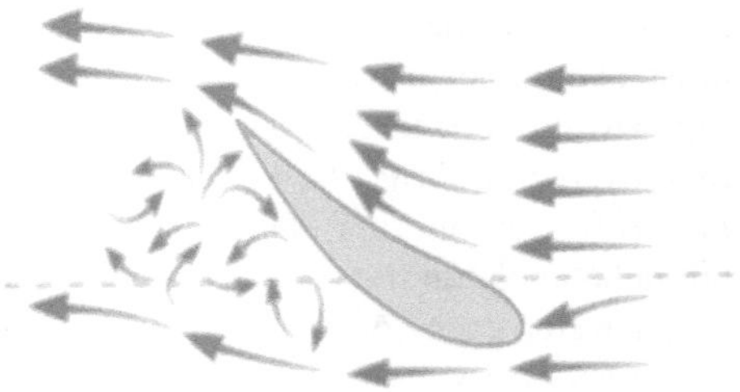

Fig. 1.5 Stall angle. The stall point occurs at about a 17 degrees angle

extended out the window. There is no airflow about the hand since the automobile is not moving. However, if the automobile is driven at 50 miles per hour, the air will flow under and over the hand at 50 miles per hour. A relative wind has been created by moving the hand through the air. Relative wind flows in the opposite direction that the hand is moving. The velocity of airflow around the hand in motion is the hand's airspeed." [3]

For Canada geese, the additional lift generated by the increased angle of attack does not generate a significant amount of additional drag,[4] so the overall effect is to increase the amount of air flowing over the wings. In fact, the increased angle of attack works in much the same way as the angle of attack does on a kite. This is because the air that hits the raised angle of the bird's body is deflected downward and generated an equal upward force in accordance with Newton's Third Law of Motion. For a time, it was thought that this was the sole method of thrust generation, with the airstream deflected downward off the bottom of the airfoil. It is now known that this is only one aspect.

Once the Canada goose is in the air and air is moving over the bird and its wings, the second and more important Bernoulli principle comes into play. According to this principle, air flows more swiftly over the curved upper surface of an airfoil like a wing than the air flowing beneath that airfoil. The air moving at greater speed over the *cambered* (curved) upper surface produces an area of low pressure above the airfoil. This low pressure above the wing allows greater pressure to be exerted by the air beneath the wing. The greater pressure beneath the wing creates the lift the bird needs to stay aloft. This is an important component of flight because the faster the air flows above the wing, the less the pressure above the wing, and the greater the lift.

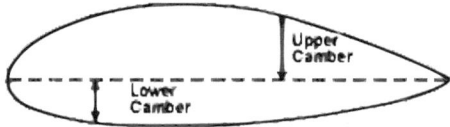

Fig. 1.6 A cambered airfoil has a curved upper surface

Flying into the wind also helps the goose to generate airflow speed over its body and wings. The body and wings act as an airfoil with a curved upper surface (*camber*) and a flatter under surface (Fig. 1.6). As the air flows up and over the top of the bird, the air must move quickly to keep up with the air passing directly beneath the wing and bird.

A goose "runs" across the surface of the water to generate the speed necessary to bring itself aloft. With a combination of flapping, gliding, running, and jumping, the Canada goose will attain the equilibrium point necessary to bring it aloft. But generating lift is one thing. Overcoming gravity to stay aloft is another.

It is not enough that the Canada goose generates lift. It must generate enough lift to overcome the force of the Earth's attraction for an object of its mass. Because of this, the exact amount of lift needed varies by bird with the weight of the bird. One thing that can be said for all, however, is that the bird—or any object working to achieve flight—will not be able to rise and remain aloft until the lift it generates is greater than the force of gravity pushing down upon it.

But a bird is not the same as a helium balloon on a string. Simply rising into the air is not the ultimate goal. The goose seeks forward motion. To accomplish this, it must overcome the resistance (friction) it encounters as it moves through the air and generate sufficient thrust. The wing makes this possible.

All bird wings have several structures in common. The Humerus is the bone that attaches the wing to the body at the "shoulder." It extends to what would be the elbow in a human. The Radius and Ulna bones are roughly parallel to one another. They extend from the Humerus to the "wrist." The hand and fingers of the bird comprise the portion of the wing farthest from the body of the bird. It is made up of the carpus, the metacarpus, and the phalanges. The carpus is formed of two short bones. The metacarpus consists of three long bones. The phalanges are formed of articulated bones that bear the primary feathers [5] (Fig. 1.7).

Birds have hollow bones with crisscross structures inside that give them support. This makes the job of lifting the bird less difficult than if the bones were solid like the bones in a human. However, the hollow bones and the bones making up a bird's wing are not the deciding factors in performance. It is the feathers on the wing that determine the performance of the wing.

All birds have four types of feathers on their wings. The different types of feathers support different functions. For instance, the *scapular feathers* "overlay the wing feathers on the back when the wing is folded to the body. They provide a streamlined transition in the aerodynamic contour of the bird between body and wings." [6] The *secondary feathers* are associated with the radius and ulna bones. "The cross-section of this portion of the wing creates the airfoil that provides lift for

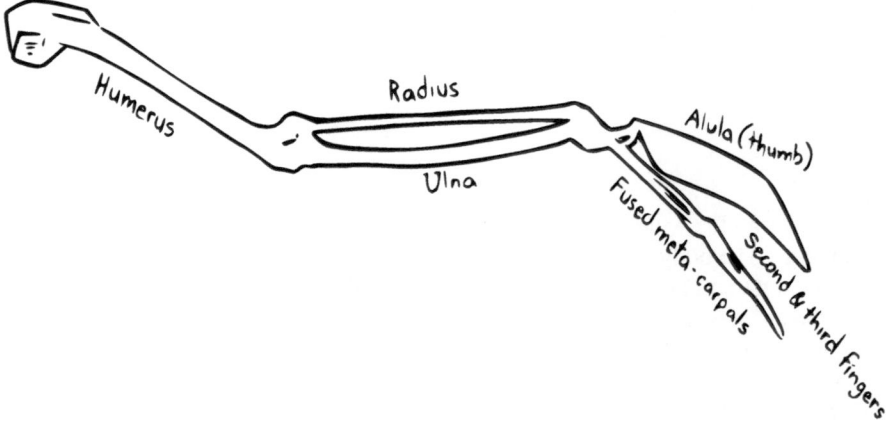

Fig. 1.7 Bird wing skeleton. All bird wings have several structures in common

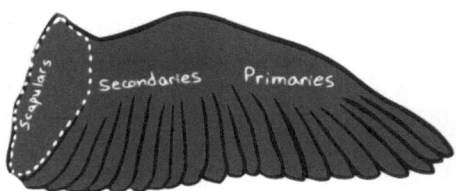

Fig. 1.8 Bird wing feathers named. Each type of feather serves a different function. The exact number of each type of feather depends upon the type of bird

a bird in flight. The number of secondary feathers varies among different species."[7] "The *primary feathers* are attached to two fused bones that correspond to the index and middle fingers of a human hand. Birds have nine to twelve primary feathers attached to the back edge of these fused 'finger bones.' They provide thrust for moving the air during lapping or hovering flight." The *alula* feathers are the fourth type of feathers. They are located on the leading edge on top of the wrist joint. "The alula comes into play when a bird lands at an angle of attack exceeding 16 degrees. When the alula is raised, it becomes a 'slat' that forces an intense airstream along the top surface of the wing. This prevents stalling as the bird's forward speed decreases and as the angle of attack increases even more."[8] The exact number of each type of feather varies with the type of bird (Fig. 1.8).

For the Canada goose, the ability to overcome resistance and produce sufficient thrust is vital if it is to make its long migratory flight. Resistance comes from many sources. It comes from the air as it moves along the body of the bird. It comes as the bird first heads into the wind and must make its way through the wind. And it comes from any part of the bird projecting away from the body of the bird, including the wings. Resistance is the effect of friction. If there were no resistance, the bird would

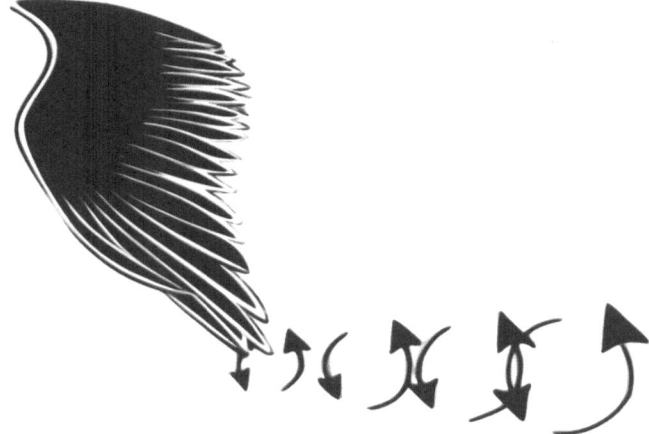

Fig. 1.9 The wingtip vortex of one bird creates an updraft for the bird behind it

fly forever without slowing down. Because of friction, the bird must continuously generate a force to propel it in the direction it wishes to proceed. Thrust provides this forward motion.

Thrust comes from the downward movement of the primary wing feathers. The force of this movement overcomes the force of drag, which is caused by the shape and contour of the bird and its wings.[9] For a Canada goose, the secondary feathers are held in a relatively horizontal position to create lift even as the primaries flap up and down to generate thrust.[10] The net result is greater lift.

When in v-formation, a strong, healthy bird is in the lead. This lead bird must be strong because it encounters the greatest *resistance*—the greatest *drag*—as it is the first to disturb the air. This disruption results in *turbulence*, the disruption of the streamlines of air. The turbulence lessens along the length of each side of the v-, requiring less effort for the birds further back to maintain their positions. There is also a positive effect from updraft of the wingtip vortex generated by the bird directly in front of each bird (Fig. 1.9). It gives each successive bird an easier ride through a phenomenon known as *drafting* as the birds take advantage of the air behind the bird in front of them. This air moves at the speed of the bird in front of them, rather than at the speed of the fluid moving around it. When the formation reaches the birds at the end of each side, there is increased turbulence as the air flows behind them and moves to resume an unimpeded flow. Still, the birds on each side at the end of the v- do not encounter as much resistance as the leading bird. The leading bird will stay in the lead until fatigued. Then it will drop back in the order and let another bird take its place.

Since fluids are said to flow *continuously*, there are no gaps in the flow. The moving air generated by the thrust of the Canada goose's wings does not have empty spaces where there is no moving air. This is because the molecules in a fluid are in close relation to one another and are part of a *continuum*. As a continuum, the mass

Fig. 1.10 A laminar flow
runs or moves smoothly, with
unbroken continuity

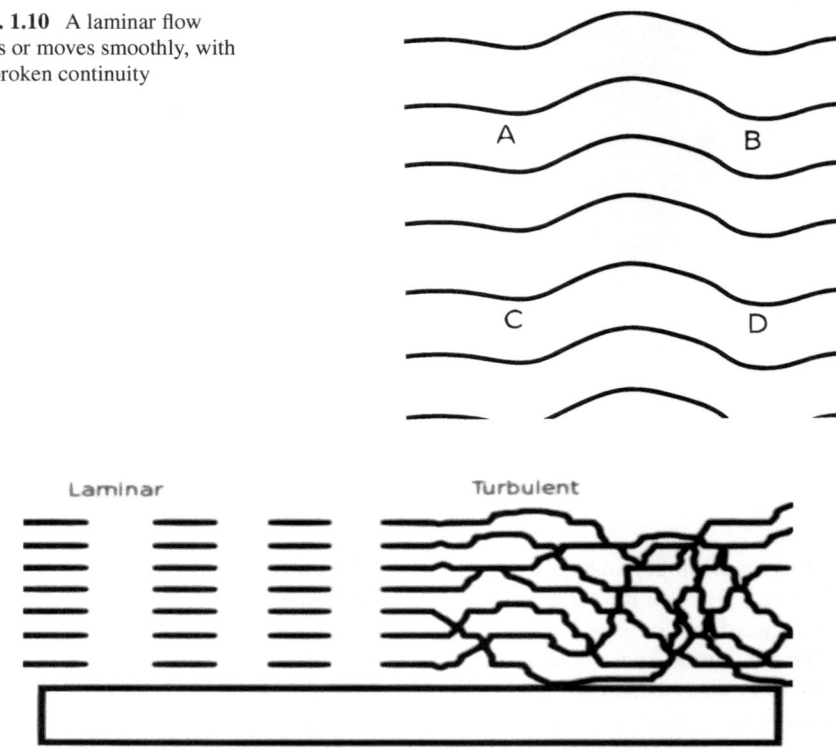

Fig. 1.11 A turbulent flow has eddies or swirls as a result of shear stress

of a fluid can be completely accounted for at different points in the fluid, and a fluid
will move or run smoothly with unbroken continuity in what Scottish civil engineer
William John Macquorn Rankine first termed a *streamline* in a paper he published
in 1871 (Fig. 1.10).

In the streamline, a continuous series of particles follow each other in an orderly
fashion in parallel with other streamlines. For the contemporary imagination, it may
be helpful to think of water having layers and flowing like parallel sets of bytes of
information with eight bits traveling alongside one another and each bit following
the one before it. In an *ideal fluid* (which does not actually exist) each streamline
would maintain its position unchanged in a steady current. In a *real fluid* there will
be events that interrupt the steady flow of the fluid.

Fluid flows in streamlines in an ideal fluid are in a *laminar regime*. They move
in parallel layers with no disruption between them. Since the flow is smooth and
calm in these imaginary flows, the molecules move in an orderly fashion, and the
paths followed by the molecules are called streamlines, another name for *laminar
(layered) flow* is *streamline flow*. Fluid flows in real fluids can be laminar or *turbulent* (Fig. 1.11). When the layers get mixed, they produce eddies or swirls, and the
fluid flow enters a *turbulent regime*. In a turbulent flow, the molecules do not stay in

Fig. 1.12 A bulbous bow reduces the resistance of a ship moving through the water

parallel layers like the bits in a byte of data. Instead, *shear stress* is in action as the layers slip past one another and the fluid churns and moves, resulting in whitewater and a bumpy ride for an object moving along with the fluid.

For Canada geese, the v-formation mimics a vessel moving through the water in both form and function, bringing tangible benefits to the birds composing it. It operates as a streamline with the shape of the goose's head and neck adding to the efficiency. This is due to the fact that the beak projects outward from the head and is the first part of the goose to slice into the wind. From the beak, the head rounds back and meets with the long, slender neck before joining with the body of the goose.

Ocean-going vessels have a similar structure below the water line. Known as the *bulbous bow*, it resembles a torpedo jutting out from the lowest portion of the hull (Fig. 1.12). It cuts the amount of turbulence at the bow and reduces the resistance because the water it encounters flows over the bulbous bow and down under the vessel, reducing the height of the *bow wave*—the amount of water rising up the front of the bow—in the process (Fig. 1.13). In other words, rather than having the water rise up along the unimpeded front bow of the ship, increasing resistance along the way, the resistance is reduced by having the water flow up and over the bulbous bow, leaving a smaller amount of water to rise to a considerably lower height up the front of the bow. The shape of the beak and head of Canada geese reduces the amount of resistance encountered in flight in a similar fashion. If the geese were not shaped in a way that is aerodynamically favorable, they would encounter even greater resistance as they moved forward.

Canada geese are not the only birds subject to the four forces of flight. Ocean birds are also subject to them as they ride the *ridge lifts* above the waves. Pioneering fluid dynamicist and English engineer William Froude commented on the birds' behavior while on an ocean voyage in 1878. In "On the Soaring of Birds," he observed albatrosses during small swells with light winds and wrote, "Under these conditions the birds *seemed* to soar almost *ad libitum* both in direction and in speed.

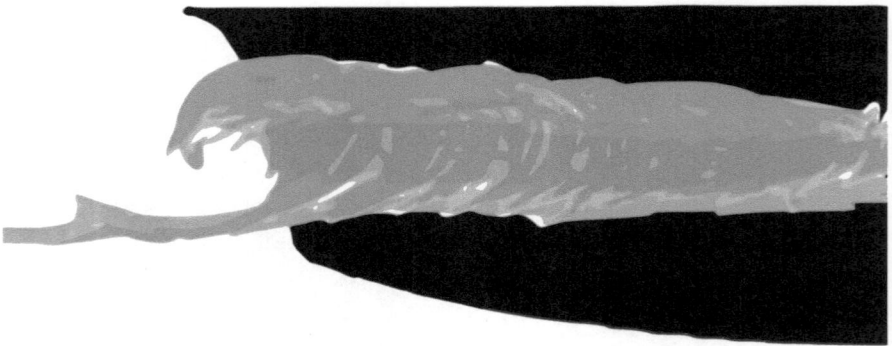

Fig. 1.13 The bow wave is the amount of water rising up the front surface of a ship

Fig. 1.14 Sea bird on ridge lift. William Froude observed sea birds riding on the ridge lifts

Now starting aloft with scarcely, if any, apparent loss of speed. Now skimming along close to the water, with the tip of one or other wing almost touching the surface for long distances, indeed now and then actually touching it."[11]

Through careful study Froude was able to conclude that the birds were taking advantage of the air that had been at the trough of each wave as it made its way up the side of the wave to the height of each wave as the wave motion progressed, in a process now know as *ridge lift* (Fig. 1.14). For the conditions he observed, Froude wrote, "it follows that all along the side of the wave at its mid-height the air must approximately be ascending at the rate of 3 feet per second, and if the bird were so to steer its course and regulate its speed as to conserve this position he would have the advantage of a virtual upward air current having that speed."[12]

Fig. 1.15 The planform is
the view you have when
looking straight down on the
object

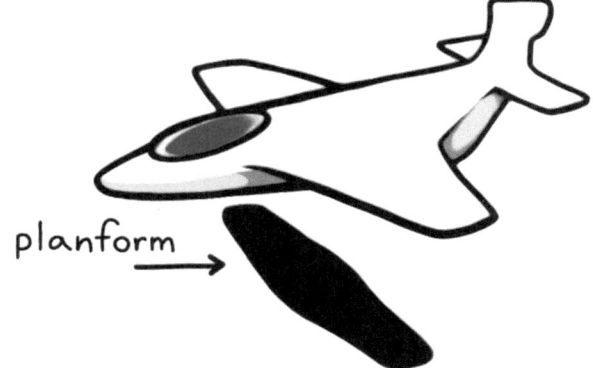

planform →

Froude continued his observations, making calculations about the effects of wing lengths, wave, and wind conditions before writing, "The voyage up from Simons Bay was delightful; for it was a glassy calm; and as there was also a tolerably pronounced swell, especially the latter part of the way, I was able (and Tower helped me) to watch the albatross's flight in a calm, with the following results:— When flying high they had to flap their wings continuously, except when descending. When near the surface they 'skimmed' occasionally, and, as far as we could distinguish, they did this only when traversing a region over an ascending wave slope. Very often this was conspicuous."[13]

The albatross is able to achieve this type of flight because "the albatross has a relatively heavy body considering the amount of wing area that generates lift. Its high wing loading would suggest that the large bird must expend a significant amount of energy in flight. However, the high wind speeds that characterize its environment compensate for the small surface area of its wings. The wind helps it maintain soaring flight for long distances with minimal energy expenditure... It also has the highest aspect ratio of all birds. This is an advantage for soaring and gliding in the wind because the long wings generate extra lift."[14]

What Froude correctly noted was that the albatross was using its wings to glide along the ridge lift generated as the air moved up the side of the wave at the wave peaks. In this way the bird used aerodynamic principles to minimize the energy expended to stay aloft. The wings of the albatross are shaped so as to make it easy for the albatross to move in this way. This is because of the *aspect ratio* of its wing.

Defined as the ratio of the length to the chord (breadth), a long, narrow wing like that of the albatross has a high aspect ratio. A short, stubby wing has a low aspect ratio. The chord of a wing may vary over the length of a wing, so the aspect ratio incorporates this into the calculation. The *planform* of an object like a bird is the view you have when looking straight down on it or its shadow (Fig. 1.15). The albatross "has long, narrow, pointed wings adapted for flight in windy oceanic environments. The constant wind provides its heavy body with enough lift to glide above the waves searching for food for thousands of miles. Pointed wingtips minimize air turbulence and diminish induced drag. This increases the efficiency of its soaring flight."[15]

Fig. 1.16 Birds ride thermals
on warm, sunny days

The albatross fascinated William Froude. The turkey vulture, or Turkey buzzard as it was known at the time, fascinated Wilbur Wright. He watched the birds sustain stable flight day after day—something that had thus far eluded all human attempts. Wilbur noticed that the birds retained their lateral balance by a light twisting of their wing tips rather than an active flapping of their wings. This observation led to what the Wright brothers termed "wing warping." Once the brothers came up with their own method of wing warping, they were well on their way to sustained manned flight. In fact, their tests of their wing warping solution proved that wing warping, in combination with the coordinated movement of the tail, would solve the vexing problems of control during flight.[16]

The Turkey vultures were easy for the Wrights to observe because they spent hours riding the lifts. However, the lifts these large land birds ride are thermal in nature. Rather than being generated by wind or air moving up the side of a wave, these lifts are the result of rising columns of warm air. These columns or pockets of warm air are in turn the result of uneven heating of the ground surface (Fig. 1.16).

Thermals rise above surfaces that absorb the heat and as a result are warmer than the surrounding area. This is why you're likely to find thermals over rocky surfaces more often than over lushly vegetated ground. The sunnier the day, the better the thermals. In fact, the best conditions for thermals occur on a warm sunny day following a somewhat cool night. When you spot cumulus clouds, keep your eyes open. You're bound to spot birds gliding on the strong thermals.

Turkey vultures take advantage of thermal lifts with their dihedral wings (Fig. 1.17). As they soar, they incline their wings slightly upward. This posture results in a distinctive shallow v-shape of the wings held above the back, creating what is known as a dihedral angle. This angle provides a significant flight benefit. Whenever a wind gust causes one of the vulture's wings to tip downward, that wing generates more lift because more lift is generated along the greater length of the horizontal wing than by the wing that is tilted upward. The net effect is to keep the bird in a steady position. It's no wonder these were the birds the Wright brothers observed for insights into stable flight.

Not all birds use their aerodynamic prowess to stay aloft. Peregrine falcons in particular deserves some recognition for using aerodynamic principles to their

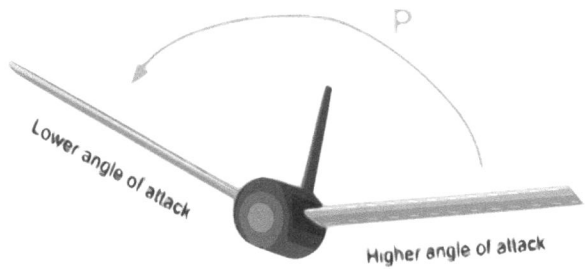

Fig. 1.17 Dihedral wing. Wings held at a dihedral angle provide a significant flight benefit

advantage in an unusual way. They first ride the thermals up to a significant height where they maintain a glide with wings flat. If done properly, the ascent can be made with very little effort on their part. Once they've achieved their optimal height and have honed in on their prey, "the bird partially folds back its wings, reducing profile drag and allowing it to accelerate. With the wings folded back, the center of gravity moves backward and the falcon is inclined downward. The tail also cocks downward, to help adjust to the downward angle of flight."

"As the falcon dives, the primaries are progressively folded straight back and held tighter to the body. This further reduces the drag. The bird shifts the center of gravity farther back and increases airspeed. Minor flicks and twists of the primaries and tail provide thrust and adjustments to the trajectory as the plummeting falcon tracks its prey like a guided missile. The peregrine becomes a feathered bullet dropping out of the sky at nearly two hundred miles per hour."[17] The Peregrine falcon captures its prey midair. It will usually strike its prey before actively grabbing it with its talons. The force of the hit can be sufficient to knock the head off the targeted animal. The ratio of dives to kills is not impressive but the successful use of aerodynamic forces does make the peregrine falcon in a dive one of the fastest animals on the planet.

Aerodynamic principles also protect the lungs of the falcon during its dive. At 200 mph, the force of the air entering the lungs would cause the lungs to explode under the pressure. The falcon survives unscathed because of baffles within its nostrils. These baffles regulate the amount of air entering the nasal cavity, and thus the lungs, during the dive. They also serve to deflect the air, which has the additional effect of allowing the bird to proceed without slowing down as the upward force of the air acts against the bird.

The tiniest birds on the planet are also worthy of consideration. Until 2005, it was widely assumed that hummingbirds used the same mechanism of flight as insects. While it's true that they do use some of the aerodynamic tricks of flying insects, wind tunnel tests prove that this initial assumption was incorrect.

Researchers knew that because a flying insect has wings that are almost flat, it employed two mirror-image halfstrokes as the wing moves back and forth in a figure-eight pattern to attain lift. This movement produces nearly equal lift during both the upstroke and the downstroke. Because they were small and exhibited flight patterns and hovering ability like that of insects, researchers assumed the hummingbird used similar strokes until they discovered their error. Although the hummingbird moves its wings in a similar pattern and can invert its wings—turn them upside down during the upstroke—to a greater extent than other birds, the figure-eight pattern of flying used by flying insets produces nearly equal lift during the upstroke and the downstroke, while hummingbird flight produces 25 percent of the lift on the upstroke and 75 percent of the lift on the downstroke. This is possible because hummingbirds have a high angle of attack and also hold their wings in a tucked position, rather than flapping them from the shoulder. This results in wing beats that are in the direction of front to back rather than up and down.[18]

Researchers were able to observe all of this because of new technology that included digital particle imaging velocimitry.[19] "This system atomizes olive oil into microscopic droplets that are so light they move instantly with the slightest movement of air—and a pulsing laser than illuminates the droplets for incredibly short periods of time that can be captured by cameras, and illustrate exactly the swirling movement of air left by a hummingbird's wings."[20]

"What a hummingbird has done is take the body and most of the limitations of the bird, but tweaked it a little and used some of the aerodynamic tricks of an insect to gain a hovering ability. They make use of what is, in other birds, an aerodynamically wasted upstroke. Coupled with taking advantage of leading edge vortices—which you can only produce to substantial effect if you're small—and viola, you're hovering for as long as you want."[21]

This ability to vary the surface of the wing in use, thereby altering the characteristics of the forces produced by the wing, as well as presenting a change in the aspect ratio, requires a lot of energy. Nectar provides the bulk of that but once again the hummingbird uses a mechanism that is not common to birds. To power its wings, it is currently thought the hummingbird uses a mechanism in common with flying insects. It's believed that they use kinetic energy in a form akin to a nerve impulse to power some of the wing movements. This requires less energy overall than would flapping their wings in the conventional manner.

By employing adaptations similar to those of flying insects to the flight mechanism of a small bird, hummingbirds have evolved into efficient hovering machines. They are also the only bird that can fly backwards. You may spot a hummingbird as it hovers by flowers with nectar that contains 10 percent or more sugar. If not, there's a chance you may be alerted to its presence by the hum produced by its wings beating at 12–80 times per second, depending upon the species.

Bats

Bats are the only mammals on Earth that fly under their own power. They have also developed a unique form of flight and are remarkably agile flyers that can turn 180-degrees at full speed in three strokes. They can do this because they have exceptionally flexible wings that enable them to take full advantage of aerodynamic possibilities while underway.

Fig. 1.18 A bat's wing resemble a human hand with a patagium stretched taut

Bat wings do not have the same properties as bird wings. It's true that bat wings are connected at the "shoulder" by the Humerus, at the point where muscles move the wing. It's also true that the Radius and the Ulna extend from the end of the Humerus. It is the "fingers" of the bat that differ significantly from the "fingers" of a bird. The bat's "thumb" extends from the upper point of the wing and the bat uses this to climb. The rest of the fingers are connected by a thin membrane of skin known as the *patagium* that covers the entire wing. Because of the malleability of the patagium and the functioning of the long, slender fingers, bat wings are shaped like human hands. When they fly, bats grasp the air and the patagium stretches taut like a sail in a breeze. The "fingers" and muscles in the wing also allow the wing to change shape while in flight. No other flying creature has this capability (Fig. 1.18).

Bats are able to maneuver their wings in a number of unique ways because "bat wings are made of quite flexible bones supporting very compliant and anisotropic wing membranes, and possess many more independently controllable joints than those of other animals."[22] Bats have upstrokes and downstrokes, similar to other flying creatures. However, not only do bat wings move up and down, the bones of the wing deform during the wingbeat cycle and throughout the entire wing. "There is noticeable difference among distinct regions of the wing, with the bones of the third digit undergoing much larger deformations than the bones of the fifth digit, oriented in the chord-wise direction. For most bones, the magnitude of deformation varies with speed; in the metacarpals and phalanges, strain magnitudes appear to decrease with increasing speed."[23]

When high-speed photos are taken of bats flying in a wind tunnel, it's possible to see the wake, lift, and thrust of their movement. While a large bird like the Canada goose achieves flight through the use of a largely inflexible wing and hummingbirds achieve flight by inverting their wings, bats maneuver their wings in ways that are impossible for either the bird or the hummingbird.

Flying Insects

Flying insect flight is also subject to aerodynamic forces. Insect wings are different from both birds and bats—insect wings are thin membranes supported by blood-filled veins—but these tiny flyers experience the same aerodynamic constraints as their larger counterparts. Because their body mass is so large in comparison to their wingspan, it's long puzzled scientists how insects get off the ground at all, let alone at such incredible speeds. Even after years of study, the exact mechanism of their flight is not fully understood, but close study of tethered flight reveals that insects can use their wings at more than one *aspect* by presenting different surfaces to the flight. They not only "flap" or "beat" with their wings but can twist and move them to form a figure-eight pattern that increases lift dramatically and sends the insect soaring. When an insect wants to hover, a different, flatter wing motion is used to keep the insect aloft and in place.

The ability to alter the aspect of the wing is the key to staying aloft. It results in some peculiar flight patterns as they employ this capability, but the net result is that the insect remains aloft and attains forward motion—often at a high rate of speed—through various flying patterns that provide amounts of thrust that are sufficient for their needs.

Part of the energy that provides the motion of the wing itself is now thought to be kinetic in nature, with some of the energy being returned to the insect as a type of nerve impulse. It is thought that this conservation and recycling of energy is what allows insects to beat their wings at incredible rates per second without exhausting themselves. Conventional sources of energy are also used for flight, which is produced in two ways. There are muscles that are directly attached to the wings and muscles that are indirectly attached to the wings.

With muscles attached directly to the wings, there is one pair of muscles for the upstroke of the wings and another for the downstroke. As the muscles contract, the wings attached to them are moved upward or downward. Coordination of these strokes is provided by the insect's brain. It requires thought to synchronize the wings and initiate the flapping motion. This results in relatively slower flight. For indirect muscle systems in insects, the muscles run the length of the thorax. As these muscles contract and relax, they force the *tergum* (dorsal portion of an anthropod segment other than the head) up or allow it to relax back into the resting position. The movement of the tergum results in the movement of the wings. Because there is no thought or coordination of wings involved in this method, wings can beat at a far more rapid rate. The wings are synchronized, too, because the tergum causes both wings to move at the same time. The net result is rapid, synchronized flight.

Many tiny insects such as ballooning spiders, mites, and some caterpillars "fly" without the use of wings. They accomplish this by employing gossamer threads to carry them aloft in much the same way as a cartoon character being carried away by a bundle of balloons. The tiny spiders climb to a high point, stand on their toes, and release several strands of silk. These strands form a parachute that allows the spider to be carried by the wind. Depending upon the wind velocity, the spider will be carried a short distance—or much farther. It's even been observed that tiny insects riding the air have been carried by airstreams and other currents. These rides have resulted in travels of significant distance.

Seeds

Tiny insects are not alone in being dispersed by the wind. Dispersal of seeds by the wind is yet another instance of aerodynamics at work in nature. From the fine dust of pollen to the robust tumbleweed, seed-bearing plants have adapted to take advantage of the wind through a variety of passive mechanisms.

Gliding seeds such as the Asian climbing gourd *Alsomitra macrocarpa* have seeds that are carried aloft on a single wing that mimics the shape of a stealth bomber. The wingspan is 5 inches and the wing glides in wide circles through the air of the rain forest.[24] Sycamores and maples are two trees with seeds that are carried on fixed uni-wings (Fig. 1.19). Perhaps as a kid, you opened one end of one of the wings and wore it on your nose. Perhaps you watched the wings acting as gliders on the wind. The seeds spiral down in 20-foot circles. On a windy day, they can go even farther,[25] before the seed reaches the ground and begins the work of germination.

Other seeds use a parachute model. Dandelions fall into this category (Fig. 1.20). If you find a dandelion at the optimal moment, it is a mountain of fluff on a stem. Blow the fluff and off it goes. But it's not just fluff. It's actually a combination of a parachute and a tiny seed carried at the bottom. Very lightweight, the fluff and seed are easily carried aloft by the breeze. The parachute carries the seed along until aerodynamic conditions cause it to land. After it lands, the seed breaks away and germinates if conditions are favorable.

Helicopters, or whirlybirds, have a seed with a stiff stalk ending in upward-tilted wings. The pitch of the wings causes the seed to spin as it falls. The seed can be carried a considerable distance if the wind is right. You'll know one of these when you spot it; the motion is similar to when a helicopter descends slowly after a power loss.

Fig. 1.19 A uni-wing seed dispersal mechanism mimics a stealth bomber

Fig. 1.20 Parachute seed
dispersal. Dandelions use a
parachute model of seed
dispersal

Fig. 1.21 Flutter seed
dispersal. Single-winged
helicopter seeds use a flutter
seed dispersal mechanism

One final method of seed dispersal by wind is that of single-winged helicopter
seeds. Also known as flutters or spinners, these papery wings encompass the entire
seed (Fig. 1.21). When they are dispersed by the wind, they flutter or spin, depending upon the wind velocity and size of the particular seed.

The specific type of movement depends upon the properties of the dispersal
mechanism, but the wind carries all of these seeds a distance from the trees or plants
that generated them.

Land Animals

Birds, bats, seeds, and insects are not the only creatures in the air. Larger land animals use aerodynamic principles to their advantage, too. The flying squirrel is a prime example. These nocturnal mammals have a flap of skin known as the *patagium* attached at the wrists and ankles. When a squirrel is ready to move to a new tree, it catches enough wind beneath the patagium "to allow an angular descent to the next tree. At the last second, the squirrel can throw up its tail and forelimbs and rise or turn before dropping onto the trunk…"[26]

Flying squirrels do not actually fly, but rather attain forward motion through a combination of leaping and floating. "A normal flight for them might be 30 to 50 feet, although they have been known to travel farther than 250 feet, and they glide with a vertical drop of one foot for every three feet or so of horizontal glide"[27] (Fig. 1.22). The flying squirrel cannot take off from the ground like a bird. All it can do is leap from a branch or other tall object, spread its legs to expand its skin flap to act as a parachute, and slow the rate of descent as it glides to the next location. This combination of forward motion provided by the initial leap and lift provided by the "parachute" of skin supports the squirrel as it moves from one location to the next.

The prairie dog uses aerodynamic forces in another way. Because these communal animals live in the wide-open spaces with little above ground shelter, they scamper into their extensive network of underground tunnels and burrows the moment they spot a predator overhead. With so many prairie dogs residing in this enclosed

Fig. 1.22 Flying squirrels gain forward motion through the use of their patagium

Fig. 1.23 Prairie dog tunnels take advantage of the Bernoulli Principle to keep their air fresh

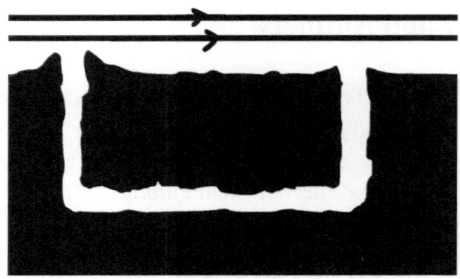

space, it would be easy for the air quality to suffer; it does not because prairie dog tunnel entrances employ Bernoulli's Principle (Fig. 1.23).

When applied to an airfoil such as an airplane wing, the principle states that the air moving up and over the upper, curved surface of the wing will flow faster than the air passing directly beneath the wing. The prairie dog tunnel entrances are not wings but they are at different heights. When the air speeds up as it passes up and over a tunnel entrance, the swiftly moving air results in a drop in pressure above that entrance. The air in the tunnel then moves rapidly through the tunnels and out the entrance with the low pressure. Fresh air immediately moves in to fill the space left by the exiting air. In this way, the air in the tunnel system remains fresh as the result of aerodynamic principles in action.

Termites are also natural engineers. It's especially important in dry, hot country for them to maintain temperatures that are tolerable for their colonies. This is possible because the termite mound itself acts as a chimney, allowing the hot air from their underground tunnels to escape. The updraft created by the rising hot air and Bernoulli effect as the airstream crosses the top of the tower, draws fresh air in through narrow tunnels that lead from the sides of the mound to the interior. The result of the aerodynamic forces at play is a ventilated environment at an ideal temperature for the termite inhabitants.

Conclusion

From skyward to earthbound, a wide variety of birds, land animals, and plants have found a way to adapt the principles of fluid dynamics to their needs. Some have used them to attain flight. Others have used them to ride on different types of updrafts or thermals. Still others have employed the principles of fluid dynamics to disperse their seeds—or their young—or to keep the air fresh in their tunnels and burrows. It's clear that fluid dynamic forces apply in a large number of processes that would not at first occur to us to involve fluids, but over time the advantage has gone to those species using fluid dynamic principles to their advantage.

Notes

1. Henderson, C. L. (2008). *Birds in flight: the art and science of how birds fly*. Minneapolis, MN, Voyageur Press.
2. Garber, S. (2011). "Centennial of Flight." *Born of Dreams - Inspired by Freedom*. Retrieved various, 2011/2012, from http://www.centennialofflight.gov.
3. (2000). "Pilotfriend." 2012, from www.pilotfriend.com/.
4. Henderson, C. L. (2008). *Birds in flight: the art and science of how birds fly*. Minneapolis, MN, Voyageur Press.
5. Alexander, D. (2002). *Nature's flyers: birds, insects, and the biomechanics of flight*. Baltimore, Johns Hopkins University Press.
6. Henderson, C. L. (2008). *Birds in flight: the art and science of how birds fly*. Minneapolis, MN, Voyageur Press.
7. Ibid.
8. Ibid.
9. Ibid.
10. Ibid.
11. Froude, W. (1878). On the soaring of birds. *The Papers of William Froude, M.A., LL.D., F.R.S. 1810–1879*. R. N. Duckworth, Captain A. D. London, The Instution of Naval Architects: 340–344.
12. Ibid.
13. Ibid.
14. Henderson, C. L. (2008). *Birds in flight: the art and science of how birds fly*. Minneapolis, MN, Voyageur Press.
15. Ibid.
16. Stimson, R. (2001). "WrightStories.com." 2012, from http://www.wrightstories.com/history.html.
17. Henderson, C. L. (2008). *Birds in flight: the art and science of how birds fly*. Minneapolis, MN, Voyageur Press.
18. Stauth, D. (2005). "Research Shows Hummingbird Flight An Evolutionary Marvel." from http://oregonstate.edu/ua/ncs/archives/2005/jun/research-shows-hummingbird-flight-evolutionary-marvel.
19. Ibid.
20. Ibid.
21. Ibid.
22. Swartz, S. M. I.-D., Jose; Riskin, Daniel K. (2007). "Wing Structure and the Aerodynamic Basis of Flight in Bats." *American Institute of Aeronautis and Astronautics*: 10.
23. Ibid.
24. "Wayne's Word." *Blowing in the Wind: Seeds & Fruits Dispersed By Wind*. 2012, from http://waynesword.palomar.edu/pifeb99.htm.
25. Loewer, H. P. (2005). *Seeds: the definitive guide to growing, history, and lore*. Portland, OR, Timber Press.
26. Resources, S. C. D. o. N. (2012). "South Carolina Wildlife." 2012, from http://www.scwildlife.com.
27. Ibid.

Chapter 2
Human Innovation

> *Any object, wholly or partially immersed in a fluid, is buoyed up by a force equal to the weight of the fluid displaced by the object.*
>
> Archimedes

In Greek mythology, Icarus takes to the sky on wings of feather and wax. All goes well until he flies too close to the sun, his wings melt, and he plunges into the sea and drowns. Despite Icarus' difficulties, flight under our own power has intrigued man throughout time. The birds make it look so easy. How can it be that this talent eludes us?

The simple answer is that birds are made to fly. From their structurally sound hollow bones to the special-purpose feathers covering their wings, birds are ideally suited to flight. Humans are not. Our bones are solid, not only adding to the mass we must lift but also putting an additional burden on the muscles we would use for flight. Because of this, the aerodynamic principles at work in the four forces of flight simply do not work in self-powered human flight to the same effect as they do in flight in birds. This hasn't kept humans from trying to achieve heavier-than-air flight under their own power, however.

The first human who fashioned wings out of palms or other plant material, lashed them to his arms, and ran as fast as he could with arms flapping, must have looked incredibly foolish as he tried to rise into the air. He was also doomed to fail since there was no possible way he could generate enough *lift* to overcome the forces of *gravity*. Likewise for the first human who lashed similarly fashioned wings to his arms and leapt off a high bluff with arms flapping. He may have looked like he was flying for a moment or two, but in actuality he was just "falling with style" the entire time. The wings he'd fashioned could not possibly generate enough lift to overcome gravity and keep him aloft. It's also doubtful that the wings he fashioned could provide *thrust*—the force needed for forward movement.

G. Hagler, *Modeling Ships and Space Craft: The Science and Art of Mastering the Oceans and Sky*, DOI 10.1007/978-1-4614-4596-8_2,
© Springer Science+Business Media, LLC 2013

Fig. 2.1 da Vinci ornithopter.
da Vinci sketched his ideas
for an ornithopter in the early
1500s

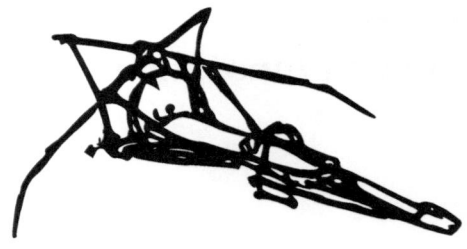

Leonardo da Vinci correctly concluded that man was not going to fly with a simple set of wings attached to his arms. In the early 1500s da Vinci sketched his ideas for an *ornithopter* (Fig. 2.1). The ornithopter closely mimicked the anatomy of a bird, and the idea was that a human would lie on the base of the ornithopter and cause the wings to flap by maneuvering a series of levers and pulleys. The model looked good, but it would not have worked. The wings simply could not generate enough lift to get the contraption off the ground, let alone sustain flight and provide the necessary thrust for forward motion.

Giovanni Battista Danti, a contemporary of da Vinci's, thought he had the solution for self-powered human flight. He glued feathers to his arms and flapped his arms up and down as he ran. His only accomplishment was to repeatedly crash onto the roof of Saint Mary's Church by Lake Trasimeno near Perugia, Italy. In the 1600s, an Italian named Paolo Guidotti built wings of whalebone, covered them with feathers, and curved them into a wing shape with the use of springs. It took a fall through a roof and a broken thigh to convince him that feathers held no magic.[1]

While it's true that bird feathers alone do not possess magic, we now know that they do play a vital role in the aerodynamic functioning of a bird's wings as a bird balances the four forces of flight: *lift, gravity, thrust, and drag*. The *scapular feathers* facilitate "a streamlined transition in the aerodynamic contour of the bird between body and wings."[2] Without this specific type of feather atop the shoulder portion of the human body, there would be protrusions and interruptions in the streamline that would create resistance and impede flight.

Likewise, without the *secondary feathers*, a true *airfoil* would not be attained because "the cross-section of this portion of the wing creates the airfoil that provides lift for a bird in flight."[3] Without an airfoil, Bernoulli's Principle would never come into play because there would be no reason for air to move quickly up and over the cambered portion of the airfoil while moving more slowly beneath the flat portion of the airfoil. As a result, *lift* would not be created.

The *primary feathers* are equally important because these are the feathers that provide the *thrust* necessary for forward motion. Similarly, the *alula feathers* are essential to flight because they work to keep the bird in flight as the angle of attack increases in excess of 16 degrees and a stall results.

For a bird, all four types of feathers are essential to the production of the four forces that allow flight. *Lift* must be generated to overcome *gravity*. *Thrust* must be sufficient to overcome *drag*.[4]

For a bird, the structure of the wing alone is not enough; the feathers must also be present. For a human, neither the structure nor the feathers, or even the combination of the two, are enough to do the job. We are anatomically incapable of flight as achieved by the birds. We simply cannot generate the four necessary forces on our own. Unfortunately, it would take many more failed attempts and an additional 100 years for this to be fully understood.

The Kite

Through all mankind's experimentation with flight, the lowly kite has served as both entertainment and a testing mechanism for aerodynamically sound design. The Wrights used kites to test their wing warping theory. They also used kites to test the design of their gliders and powered airplanes by flying them unmanned, as kites, from 1900 to 1903. The kite is an excellent choice for this because it is subject to the four forces of flight and provides a straightforward way for observing the results of changes in those forces.

Because the weight of a kite is negligible, generating sufficient lift to get it aloft is not that difficult. Neither is it difficult to keep it aloft. Once in the air, it's possible to vary the *aspect angle*—the angle of the kite to the wind (Fig. 2.2) and observe the effects. In fact, since a kite is a flat surface, rather than a cambered one, the lift on a kite is largely generated by the aspect angle, and to a limited amount, by the Bernoulli effect. It's simple to understand when expressed in terms of Newton's Third Law. This law states that the mutual forces of action and reaction between two bodies are equal, opposite, and collinear. So, when the air strikes the face of the kite that is attached to the string and at an angle to the ground, that air is deflected downward. As it is pushed downward, it in return pushes back against the kite, moving it upward. The Bernoulli effect has nothing to do with this aspect of lift, since that effect is created by the air as it passes over the top of the kite while passing beneath the kite.

For a long time, theorists believed that all of the lift generated by an object was generated by the action of Newton's Third Law. This is not the case in most instances; certainly not with the configuration of the modern fixed-wing aircraft we know today. As for the kite, the shape of the kite doesn't matter. The forces acting on it are the same no matter what the kite looks like. It may be necessary to control the kite with some variations due to the design of the kite but the *thrust* necessary for forward movement is supplied by the tension in the line that is attached to the kite. With the wind blowing parallel to the ground, *drag* is in the direction of the wind. *Lift* is perpendicular to the wind. Both of these forces act on the *center of pressure* of the kite—the spot where lift, drag, and gravity combine. This center of pressure is what makes the kite fly straight.

Lift in a kite is generated by the deflection of the wind by the kite. The wind strikes the bottom surface of the kite and is deflected down at the angle of attack. In accordance with Newton's Third Law, the kite moves upward because the downward

Kites

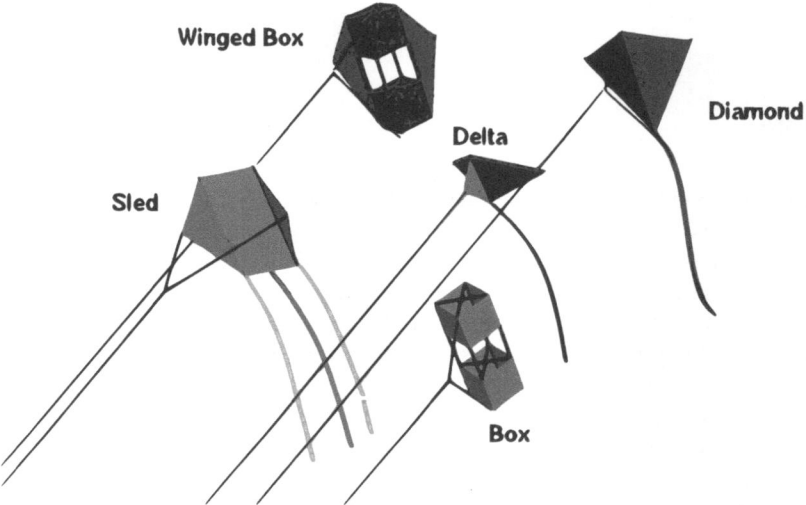

Fig. 2.2 Angle of attack of a kite. The aspect angle is the angle of the kite to the wind

action of the wind has an equal and opposite reaction in the opposite, upward direction. The Bernoulli Principle also applies to kites, though not to the extent it would in a true cambered airfoil. As the air flows up and over the kite, the pressure is less than the pressure flowing beneath the kite. As a result of the low pressure above the kite, the kite rises. The amount of air flowing up and over the kite also depends upon the *aspect ratio* of the kite—the angle of the kite to the wind.

The tail on a kite adds stability and balance. It also acts as *drag*—an increase in the resistance the kite must overcome to stay aloft. Because of the effect of drag, a kite with a tail won't fly as high as a kite without a tail. The trade off in balance and stability comes at the expense of height.

If a standard kite is the equivalent of a bird in flight, a delta-wing stunt kite is the bat of the kite universe. Stunt kites are still subject to the four forces of flight, Newton's Third Law, and the Bernoulli effect, but the way they react differs greatly from a traditional kite. The secret to the stunt kites performance is symmetry of the kite and the ability to control each wing rather than one wing alone (Fig. 2.3). To allow control over both sides of the wing, a stunt kite has two lines that are used to fly the kite. These lines are precut to the optimal length for the performance of the specific kite. This is because the goal of a stunt kite is not simply to rise as high as possible; it's to perform a variety of maneuvers. The entire length of line is let out and the kite is flown with all the line out at all times.

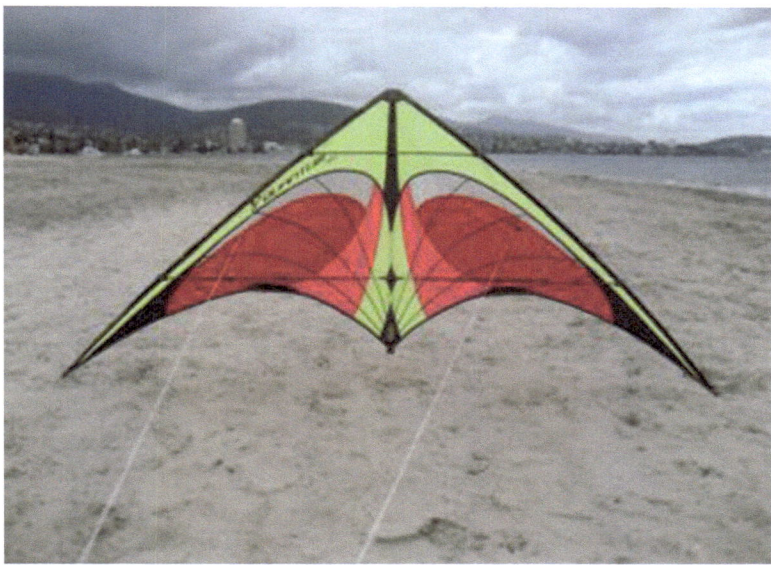

Fig. 2.3 Stunt kite. The secret to the performance of a stunt kite is the ability to control each wing rather than one wing alone. *Source*: Retrieved from http://en.wikipedia.org/w/index. php?title=File:Steve_Hobart_Sport_Kite.jpg&oldid=483266052

The ability to regulate the thrust in two locations versus only one is a key component of the aerodynamic performance of the stunt kite. In the same way that a bat can change the shape of its wings while in flight, a stunt kite can have changes in the aspect ratio and angle of each wing individually, giving them a broader range of movement while in flight.

The strongest wind will be directly in front of the person holding the kite strings. Because of this, most maneuvering will be done to one side or the other. However, in the same way a dihedral wing structure like that of the Turkey vulture corrects for stability, a stunt kite will remain stable as it moves in response to a tug on one string.

Sports

The spirit that led early aviators to take the sky lives on. Today man employs aerodynamic forces to glide in a variety of manners that include gliders, hang gliders, and parasails. With a parasail, rather than begin from a high point and glide to the earth, a boat is used to tow the person wearing the parasail into the wind like a giant kite. When sufficient lift is generated, the person rises into the air. He then glides with a huge air-filled airfoil attached, landing safely after the boat slows and ceases to generate lift (Fig. 2.4).

Fig. 2.4 The person wearing the parasail is lifted into the air as if by a giant kite

Modern-day hang gliders control their flights by hanging beneath the wing in a horizontal position. This gives the pilot greater control over the center of gravity of the glider, as well as the ability to make a wider variety of changes in position during flight. Changes in the angle of attack are made by pushing on bar that runs perpendicular to the flyer, beneath the planform of the wing.

In keeping with his quest to move, unencumbered, through the sky, Patrick de Gayardon developed the modern wingsuit during the 1990s. These suits purposefully take advantage of the principles of aerodynamics to enable a human being to leap from a plane or sufficiently high point and fall in a controlled and sustained glide, without any external apparatus, until the point where they must open a parachute to slow sufficiently for a safe landing. They have to use a parachute because it is not possible for the flyer to slow enough to land without injury, due to stalling and falling to the ground.

When wearing a wingsuit, a human mimics a flying squirrel, with flaps of fabric between his legs as well as between his arms and body (Fig. 2.5). The flyer's entire body becomes an airfoil, controlled by the movement of different parts of the wingsuit flyer's body. Flying squirrels have one configuration of skin flaps, legs, and body. Wingsuits come in a number of different configurations that meet the specific purpose of the wearer. Some are designed to sustain the glide for as long as possible; others are designed to permit greater mobility and lift during the glide.[5] The ability to achieve different objectives is based upon the aerodynamic forces at work with a specific type of suit. If greater lift is desired, a greater camber and/or angle of attack will be important. To prolong the glide, the ability to maintain a stable path might be the overriding objective in the design of the suit. It's up to the person wearing the wingsuit (birdman) to select a suit with the characteristics required to achieve the type of flight desired.

Fig. 2.5 A wingsuit allows a human to mimic a flying squirrel

Not everyone who mimics the behavior of birds is interested in taking to the air. Tune in to the Tour de France and you'll find cyclists on the same team following close behind one another, mimicking the behavior of geese in formation. The resistance is greatest for the lead cyclist while those in the middle have to do less work to cover the same ground. After a turn at the lead, that cyclist will drop back and another will buffet the resistance until it is his time to take a "break" at the back.

As is the case of Canada geese flying in v-formation, the cyclists following the leader are taking advantage of several aerodynamic forces. They are following the cyclist in front of them and enjoying a reduction in the resistance (friction) they encounter as they proceed in a process known as *drafting*. This resistance plays a significant part in the speed a cyclist can attain. Just how great a part was illustrated by two-time Olympic cyclist John Howard when "he mounted a wind-breaking shield on the back of a race car and rode his bicycle behind it, so that he was effectively riding in zero wind. He quickly got up to such a high speed that he couldn't turn his pedals fast enough, even in his top gear. So he went home and built a special bike with enormous gears, then tried it again. Using only the power of his legs but without any air resistance to fight, he hit 152 mph. A few years later Fred Rompelberg of the Netherlands gave it a whirl and got up to 170 mph."[6] The difference between the fastest they could to without the shield and the speeds they attained when riding behind the shield can all be attributed to the effects of *resistance*, the effect of drag—or friction—generated by the air flowing around the cyclist and his bicycle (Fig. 2.6).

By riding in a single file, arms tucked and legs in rhythm, the cyclists at the Tour de France are trying to achieve a similar advantage. They are also minimizing disruption to the air as it flows around them. The *streamline*, or flow around the cyclists, will have less turbulence when the cyclists are in their tucked positions

Fig. 2.6   Bike behind barrier. Olympic cyclist John Howard reached a speed of 152.284 mph while riding behind this wind-breaking shield on July 20, 1985. *Source*: Courtesy of John Howard

than it would if the cyclists were sitting straight up, their heads at varying heights, and their elbows jutting out to the sides. Anything they can do to form an aerodynamically sound "structure" will require less effort on their part and increase the speed they can attain.

Man hasn't just turned his understanding of aerodynamic principles to the sport of cycling. He's also invested considerable energy into the advantageous use of aerodynamics in baseball. For pitchers, an understanding of aerodynamics and a variant of Bernoulli's Principle have resulted in the ability to achieve a different outcome each time the ball leaves their hand.

When a pitcher throws a new, regulation baseball, he's throwing a completely round object that is smooth except for the slightly raised stitches that hold the ball together. "The fact that a baseball has low density, meaning its weight is low for its size, increases the aerodynamic effect."[7] "It's all about the spinning which is how a pitcher puts his 'stuff on the ball': by spinning the ball in different directions as he releases it, the pitcher can throw a slider, a curveball, a cutter, or, if he manages to throw it with no spin at all, a knuckleball."[8] If the ball is a tiny, immaculate orb without a nick or mark, where does it get its aerodynamic properties? The stitches.

Because the stitches are the only raised part of the ball, a pitcher who holds the ball so that the stitches are at a specific position when he begins his pitching motion can generate a state of disequilibrium as the ball moves through the air upon release. Especially for a curve ball, the air will be flowing more quickly over the stitches, creating what is known as a *Magnus force* (Fig. 2.7). It's not the same as the Bernoulli effect but it is based on the same principle. With the Magnus force, it's the stitches on the spinning ball that force the ball to move more quickly on one side than on the other. This creates an area of low pressure on the side with the faster movement. The path of the ball will curve in that direction as a result (Fig. 2.8). It wouldn't curve if

Fig. 2.7 The Magnus Force
is the force that causes a
curve ball to curve

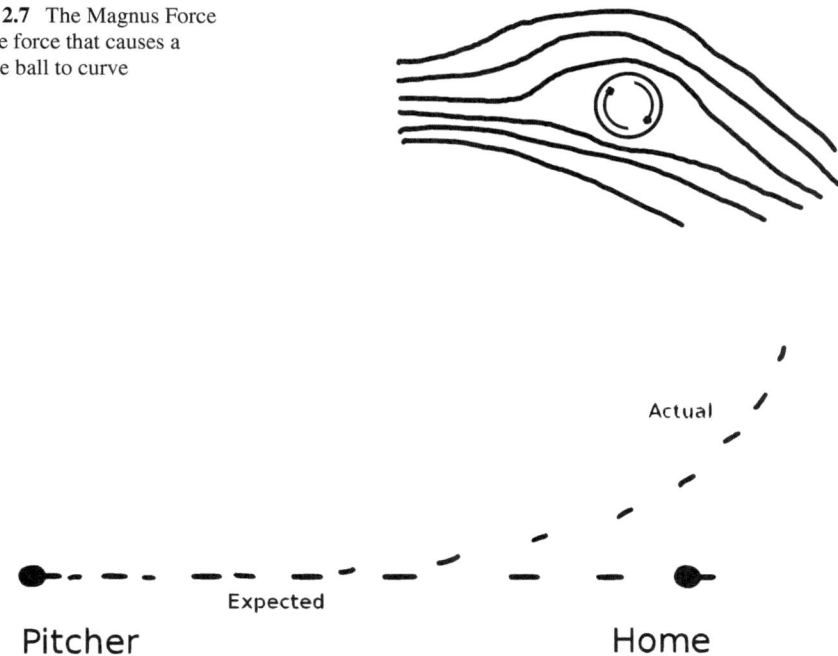

Fig. 2.8 Curveball. For years people argued about whether or not the path of a curveball really curves

the stitches weren't held in precisely the correct position and spin wasn't applied at the release. And so, the pitcher controls the flight of the ball by taking advantage of the aerodynamic properties of the raised stitches.[9] The different grips and releases result in different pitches because the aerodynamics differ with the extent of the Magnus force involved. The Magnus force is the force "directed at right angles to the direction of the air velocity and to the axis of spin."[10]

Since the batter can't see the orientation of the stitches as the pitcher releases the ball, he's left to observe the flight of the ball as it comes toward him at speeds of around 100 mph. If the pitcher "throws a 99 mph fastball, the ball is going to reach the batter in less than four tenths of a second, 395 milliseconds (ms). By comparison, it takes 400 ms—four tenths of a second—to blink your eye completely.

A lot has to happen in those 400 milliseconds. It takes the first 100 for the batter to see the ball in free flight and get an image to his brain. The brain then needs 75 ms to process the information and gauge the location and speed of the ball. In the next 25 ms—a fortieth of a second—he has to decide whether to swing, and then he's got only 25 ms more to decide if the ball is going to be high or low, inside or outside. If the decision was made to swing, another 25 ms are needed for the legs to react and begin the first motions of the swing. That leaves a grand total of 150 milliseconds for the batter to get the bat around and make contact."[11]

Fig. 2.9 America's Cup yacht. The trimaran hull on this America's Cup yacht affords a minimum of resistance

Fig. 2.10 The wetted surface of a vessel is the surface beneath the water. It is a source of resistance

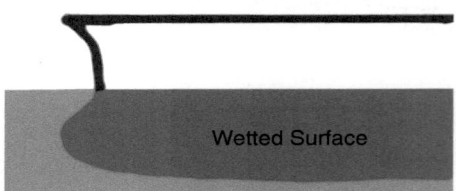

All of that is complicated enough, but "if the ball's actual path over the last fifteen feet doesn't match their mental extrapolation, the ball isn't going to end up where they think it will be."[12] And that is where the distance from the mound to the plate makes all the difference. Both are situated so that the Magnus force will cause the ball to curve, or sink, or move in an unanticipated manner within those last crucial feet, causing the batter to swing and miss.

The designs of America's Cup racers also take full advantage of hydrodynamic principles to reduce resistance and maximize speed. One way this is done is by the use of a trimaran hull (Fig. 2.9). This design minimizes the amount of the hull that forms the *wetted surface* at any given time. This is significant because the wetted surface is a major source of resistance. With anything moving through a fluid, there will be a wetted surface. This is the portion of the object that is in direct contact with the fluid. With an airplane, the entire plane is in contact with the fluid at all times. With a ship or boat, the portion of the vessel below the waterline is the only part of the vessel in direct contact with the water, while the rest of the vessel is in contact with the air (Fig. 2.10).

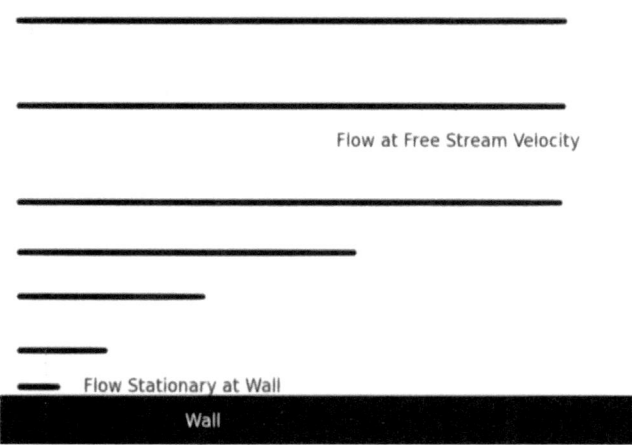

Fig. 2.11 The boundary layer is the area of greatest friction in a fluid flow

The fluid flowing past an object flows in a *streamline*. If no protrusions or other impediments are encountered, the fluid will flow smoothly, and there will be minimal turbulence. In an *ideal* (imaginary) *fluid* there would be no turbulence if there were no obstacles because the fluid in question would have no *viscosity*. Viscosity is the friction in a fluid. It determines how easily a fluid pours. Water is less viscous than honey, for example. And warm honey is less viscous than cold. When a viscous fluid flows past a wetted surface, a *boundary layer* is created. This boundary layer is an area where the forces of friction are so strong that the fluid moves very slowly, if at all. The slowest portion of the fluid slows the fluid directly beside it, and that portion in turn slows the portion beside it. The farther you move from the boundary layer, the weaker the force of friction and the more swiftly the fluid flows. At some point you will reach the portion of the flow that is unimpeded by the force of friction (Fig. 2.11).

In a vessel that rides low in the water because of a heavy load or a weighted keel that is used for balance, a significant portion of the hull is below the waterline and generating resistance that must be overcome. With a trimaran design, there is one main hull and two other hulls acting as balancing arms akin to Polynesian canoe designs. Only a small portion of the entire hull is below the water line because only the central hull is in the water at any given time. The other two hulls are never deep in the water. There is also no weighted keel required for balance, so the sailboat sits high in the water with a relatively small amount of her hull beneath the water. The combined effect of the small wetted surface and the superior balancing apparatus results in a world-class vessel capable of winning the America's Cup.

If there is an aspect of play to sport, the Frisbee flying disk is surely emblematic of it. Its concept is simple. It is shaped like the cross-section of an airfoil. It generates its own lift as it spins, allowing it to fly through the air. There are many variations on the flying disk but all owe their start to the time Walter Frederick Morrison was playing catch with his future wife during a Thanksgiving Day party in 1937.

They started out playing with the lid from a popcorn maker. "When flicked through the air rightside-up, the lid's smooth top side offered little resistance to the air passing over it, while its downturned edge created a baffle slowing the air passing beneath. The result: lift." As the game progressed, the lid got banged up and they switched to cake pans. "Stabilized by the spin imparted by a backhanded throw, the lid not only flew but also answered simple commands—depending on its angle when it left the hand, it would glide flat, curve or boomerang." [13]

They'd switched from playing with a large popcorn can lid to an empty cake pan and were still using the cake pan for their catches one day in 1938 on a beach in Santa Monica, California. A man Morrison described as a local beach bum walked up and offered them a quarter for their cake pan. "That got the wheels turning," Morrison told the Virginian-Pilot, "because you could buy a cake pan for 5 cent, and if people on the beach were willing to pay a quarter for it, well, there was a business."

Morrison sold cake plans for a while before working up a design for a flying disk toy in 1948. His sketch was of an aerodynamic refinement of the metal cake pan. Shortly after Morrison envisioned his new design, a private pilot named Kenneth Arnold described nine bright objects he'd observed near Mount Rainier in Washington state. This first report of Unidentified Flying Objects began the UFO craze in the summer of 1947. Morrison was ready.

By 1948, Morrison had a plastic flying disk to demonstrate as he continued to tweak the aerodynamic properties of his invention. "A lot if it was intuitive," said Phil Kennedy, Morrison's coauthor of their book, "Flat Flip Flies Straight: True Origins of the Frisbee." Because of his aviation experience, he [Morrison] knew what made a wing fly, and he applied that knowledge.[14]

In the time the Frisbee has been in existence everyone from the casual athlete to Bill Nye the Science Guy has weighed in on the forces at work in the flight of a Frisbee. The two main forces are gravity and air. Gravity is the force pushing down on the disk. Air is part of the force that generates upward lift for the Frisbee. The distance and direction of your Frisbee flight will depend upon the angle of release. The launch angle is the angle that exists as the person throwing the disk releases the disk. A Frisbee thrown at 180 degrees results in a straight throw. An angle greater than 180 degrees upon release will result in greater lift—an upward ride for the flying disk. Release at an angle of attack less than 180 degrees will lead the Frisbee to a meeting with the ground.

Lift is generated as the airflow over the top, curved surface of the spinning disk moves more quickly than the air flowing beneath the lower, less curved surface of the disk. The rim is an important component of the Frisbee because it is what helps to create the deep camber of the airfoil. In fact, without the rim, the angle of attack becomes the most important variable in the flight of the disk. Newton's Third Law is also in play with the Frisbee as the air pushing up on the Frisbee is met by an equal and opposite force pushing back toward the ground. This force results in additional lift.

Angular momentum is also in play while the Frisbee is in motion. It provides stability and is provided by the spin. The faster the spin, the greater the stability. This stability is essential for those trying to do tricks with the Frisbee. Of course, drag is in play, as well. On a windy day there will be more drag, or resistance, making it more difficult for the Frisbee to maintain its momentum. The angle of attack is one way to overcome the forces of resistance, too.

So what generates the power in a Frisbee toss? According to Morrison, it's all in the wrist![15]

Man on Land

Today, it is commonly accepted practice for automobiles and trucks to be designed to minimize resistance and drag. This has not always been the case. The earliest instance of the purposeful use of aerodynamics in automobile design occurred in 1935. The engineers at Chrysler, with the full support of founder Walter P. Chrysler, were determined to introduce an aerodynamically sound car to the American public. The car was wind tunnel tested and, in addition to a better ride due to changes in the overall design, boasted increased fuel economy and faster running speeds.

Testing a scale model automobile in a wind tunnel was a direct result of the work done in model basins by Froude and Taylor in the design of ships. By observing the behavior of air flow around automobile models with slight changes in design components, the design could be perfected in far less time and at a significant reduction in cost. The wind tunnel tests were an accepted method of design because of Froude's groundbreaking work with scale ship models. They were also possible because da Vinci had long ago theorized that it was not necessary to move an object to observe the effect of the wind on that object. It was possible to have the wind move past the object and record the results. The outcome would be identical.

Chrysler used the wind tunnel tests to learn that by putting the headlights flush with the grill, making the bumpers flush with the car, and having the nose of the car project slightly before the upsweep of the hood and windshield, the Airstream lines decreased resistance. In fact, "air resistance at maximum speed was reduced 44 percent and fuel economy was increased 57 percent at 80 mph. At moderate speeds the fuel economy advantage was in the area of 25 to 35 percent. Less horsepower was needed to move the car through the air, so the engine ran slower and less frictional wear resulted."[16] Despite the fact this decrease in resistance translated into greater fuel economy and the attainment of faster road speeds, the American public was not ready for the huge departure from traditional automobile designs (Fig. 2.12). They did not flock to buy the Airstream or DeSoto and ultimately, despite the improvements in performance and Chrysler's adamant defense of their designs, the company was forced to abandon these models. Today the Chrysler Airstream is recognized as an innovative design.

Man in Water

Man hasn't only looked to the skies for inspiration. Many have looked instead to the oceans. They've studied the movement of fish to see what economies man could incorporate to benefit us in our own activities. One area of intense focus on fluid dynamic principles has come from participants in the sport of elite competitive swimming.

Fig. 2.12 The Chrysler Airflow was ahead of its time. It achieved a 44 % reduction in air resistance at maximum speed and a 57 % increase in fuel economy at 80 mph

Fish are perfectly adapted to their aquatic environment. Their bodies are sleek; no bumps or lumps jut out at irregular intervals. Everything about them is designed to reduce resistance. A swimmer is subject to the same fluid dynamic forces as any other creature making its way through the water. In fact, "there are four primary types of drag that contribute to total body drag: (1) skin friction drag which is a tangential force resulting from shear stresses in the water sliding by the body, (2) pressure drag which is a perpendicular force on the body associated with the pressure of the surrounding fluid, (3) wave drag that occurs when a swimmer moves on or near the water surface, and (4) induced drag that is associated with water deflection off hydrofoil surfaces… "[17] For a swimmer going for the gold in an Olympic event, no source of resistance is too small to consider. They seek a state of minimum resistance.

"There are several ways in which a swimmer tries to overcome drag. One is to use a stroke technique that makes his body stay as high on the water as possible. The more of his body that's out of the water, the less water can hold it back. Another way is to make sure that his hands knife into the water as he reaches forward for the next stroke instead of inadvertently pushing forward, which is like stepping on a brake."[18]

Anything that reduces resistance is eagerly adopted. "There's only so much training a swimmer can do to make themselves stronger and improve their technique. That's why they look for ways to reduce drag that are 'free,' i.e., take no extra effort."[19] Bathing caps, shaven heads, shaven legs and arms on male and female swimmers—it's all done on a routine basis. Same thing with timing breaths to minimize disruption of the water surface and swimming beneath the water at the optimal depth for the optimal (allowed time) to mitigate the effects of surface disruption on performance—both matters of common practice.

Fig. 2.13 Speedo LZR Racer swimsuit image. The LZR Racer Suit changes the shape of the racers body. *Source*: Retrieved from http://www.nasa.gov/topics/technology/features/2008-0214-swimsuit.html

Since anything that will reduce the *boundary layer* and lessen *drag* is entertained as a possibility, it's not surprising that the current trend is toward long, one-piece racing suits that reduce resistance. These suits give the swimmer more in common with a fish or someone wearing a wing suit than a simple bathing suit possibly could. At the 2008 Summer Olympics in Beijing, swimmers wore a new type of swimsuit. It not only lessened resistance because of the sleek material that allowed someone wearing it to glide through the water with minimal drag, but also minimized resistance by forcing the body into an uncomfortable but aerodynamic configuration.

Speedo recognized that the human body is not tapered and sleek as an aquatic animal. They recognized that, "any time a muscle or loose section of skin bulges or shifts, it's going to block the smooth flow of water and impede the swimmer's forward motion." They designed the LZR Racer with the help of NASA. Because the human body has "momentary bulges of skin, fat, and muscle" when in motion, the Speedo LZR Racer is a full body-length swimsuit that "consists of a series of carefully shaped panels that push, squeeze, and compress the entire body into a more streamlined shape than the one he or she [the swimmer] started with." (Fig. 2.13).

"The LZR Suit holds all those bits tightly in place and stops them from sticking out into the water and increasing drag. A the same time, it changes the overall shape of the swimmer's body into a more streamlined configuration."[23] This aerodynamic perfection does not come easy. "The suit is so tight it takes half an hour—literally—to put on properly. Once it's in place, all that squeezing makes breathing more difficult, and it's so uncomfortable that the first thing wearers do when they get out of the pool is start tearing it off."[24]

Wearing this suit didn't just make the swimmer look more compact. It in fact made the swimmer's body more compact, reducing all possible sources of drag in the process. Speedo boasts that it requires 5 percent less effort to go the same speed when wearing the suit. So what was the effect of reducing resistance and changing the shape of the swimmer's body? Olympic swimmers wearing the suits attained the fastest times in the history of the Games. "When I hit the water, I feel like a rocket," says Michael Phelps, Olympic champion and one of the greatest swimmers in the history of the sport.

But how much was due to the LZR Racer and much can be attributed to the extraordinary measures taken in the creation of the pool? To further increase the speed of the pool, every technological innovation possible was used in creating the pool with the goal to reduce waves and the effect of those waves on the swimmers as they participated in their events. The desired outcome was what is commonly referred to as "fast water." The depth of the pool, the number of lanes, the gutter system, and the temperature of the water were all part of this effort, as was calculating the exact depth versus width of the pool to allow the maximum dissipation of disruption in the smallest amount of time. When the pool was complete, it was the epitome of a swimming environment designed to allow the peak performance of every athlete.

"In general, body drag for a swimmer moving on or near the water surface is 4–5 times higher than the level of drag encountered by the submerged swimmer moving at the same speed (Hertel 1966). Much of this increase in drag at the water surface is due to energy wasted in the formation waves."[25] Fish and marine mammals overcome this problem by swimming deep enough to avoid the effect of these waves. Olympic swimming rules restrict the distance over which a swimmer can proceed in this way.

As a result, "swimmers generate waves as they churn down the lane, not just at the water's surface but below it as well. These waves travel rapidly down to the bottom of the pool and then bounce, in the same way that a sound wave echoes off a wall. The returning wave creates turbulence that slows the racers down. The deeper the pool, the more these waves will be dampened on the way down and up, resulting in a smoother and therefore faster ride for the swimmer. Modern competition pools have a uniform depth of seven to nine feet."[26] The pool at the Beijing Olympics is 10 feet deep, 1.3 meters deeper than most Olympic pools. This is the optimal depth for a pool because it minimizes the effects of turbulence caused by the activity of the swimmers, yet is not so deep that their sense of vision is lost.

"Waves travel sideways, too, affecting swimmers in adjacent lanes. One way to ameliorate this effect is to make the lanes wider" but "even more important than the lane lines are the gutters at each end and along the sides of the pool."[27]

"In some high-end pools, such as the one at Beijing, there is an extra lane on each side, which remains unoccupied during a race. Its only function is to give lateral waves a chance to dissipate as they bounce."[28]

The Beijing pool also employed perforated gutters on both sides to absorb the lateral waves. The net effect, along with the temperature of the water—set at a point where it was comfortable for the swimmers yet reduced the viscosity of the water, was to create the fastest water yet at an Olympic swimming venue. Between the optimal conditions in the pool and the use of the Speedo LZR Racer swim suit, 25 world records were broken over the course of the Beijing Olympics. All of the improvements in conditions were "free." They required no extra training on the part of the athletes in the same way that shaving their heads or dolphin swimming upon initial entry to the pool brings improved performance.

It's interesting to note that the same concerns with dissipation of waves were expressed by the early designers of model basins for scale model testing. William Froude and David Taylor each took elaborate measures to ensure that the depth of the water would reduce the bounce back of turbulence from the model runs. Gutter systems were put in place on the sides of the basins to hasten the dispersion of lateral movement. These early fluid dynamicists sought to eliminate any forms of turbulence in their venues with their designs; today's swimmers seek to do the same with the design of their venues. The prize for their efforts is a model basin or pool that permits the best possible performance of the model or swimmer because of the active steps taken to reduce the resistance created by the test or event itself.

It stands to reason that if man will pursue perfection in the form they take while moving directly through the water, they'll want to utilize a design for peak performance when moving through the water in a craft. This has been precisely the case with the development of the submarine. From the earliest days of testing and innovation of the first truly practical and modern submarines at the start of the twentieth century, consideration of ways in which to minimize resistance have been of primary importance.

Submarines are essentially stealth craft. They make their way, unnoticed beneath the waves. If a submarine is "noisy," it will be easy to detect. If it is capable of running "quietly" it can enter areas at will without drawing attention. The same factors that make a submarine noisy are the things that reduce its hydrodynamic efficiency. These factors include anything that increases resistance or turbulence at the boundary layer. They also include cavitation or bubbles around the propeller action that disrupt the water at the prop. A submarine that can flow with a sound footprint similar to that of a shark is one that is making the best use of its propulsive power: It is running with maximum efficiency.

Since a submarine moves through the water in a manner similar to the manner in which a bird makes its way through the sky, the four forces of flight are at play in the design of these craft. The shape is important because it influences the body drag of the vessel. Streamlines are in effect for a submarine, just as they are for any craft moving through a fluid. Bernoulli's principle is also in play. This will influence the amount of lift the sub can generate.

Skin friction is another consideration. The more attention paid to the laminar and turbulent flows around the vessel, the better. This attention will result in an optimal length to go with the optimal shape, resulting in increased quite and efficiency. Appendages must also be designed for least resistance. By carefully studying the effects of hydrodynamic forces on submarine designs in model basins and in full-sized craft, submarine design has resulted in vessels that are not only efficient but also nearly silent.

Conclusion

From the time of the ancient Greeks to the time of the first submariners, humans have set their sights on moving through the air like a bird and the oceans like a fish. The result has not been the unencumbered movement enjoyed by these animals in nature, but it has been sufficient to bring humans eye to eye with the objects of their fascination in their own environment.

Notes

1. Fuller, J. (1998–2012). "How Stuff Works." *Top 10 Bungled Attempts at One-Person Flight*. 2012, from http://science.howstuffworks.com/transport/flight/classic/ten-bungled-flight-attempt3.htm.
2. Henderson, C. L. (2008). *Birds in flight : the art and science of how birds fly*. Minneapolis, MN, Voyageur Press.
3. Ibid.
4. Ibid.
5. "Birdman."
6. Brenkus, J. (2010). *The perfection point*. New York, NYEnfield, Harper ; Publishers Group UK distributor.
7. Ibid.
8. Ibid.
9. Adair, R. K. (1994). *The physics of baseball*. New York, HarperPerennial.
10. Ibid.
11. Brenkus, J. (2010). *The perfection point*. New York, NY Enfield, Harper ; Publishers Group UK distributor.
12. Ibid.
13. Swift, E. (2007). the man with the original plan. *The Virginian-Pilot*, Landmark Communications Inc.
14. Ibid.
15. Ibid.
16. Butler, D. Adventures in Airflow: Part II. *Cars & Parts*: 8.
17. Perrin, W. F., B. G. Würsig, et al. (2009). *Encyclopedia of marine mammals*. San Diego, Academic Press.
18. Brenkus, J. (2010). *The perfection point*. New York, NY Enfield, Harper; Publishers Group UK distributor.
19. Ibid.
20. Ibid.

21. Ibid.
22. Ibid.
23. Ibid.
24. Ibid.
25. Perrin, W. F., B. G. Würsig, et al. (2009). *Encyclopedia of marine mammals*. San Diego, Academic Press.
26. Brenkus, J. (2010). *The perfection point*. New York, NY Enfield, Harper; Publishers Group UK distributor.
27. Ibid.
28. Ibid.

Chapter 3
Aquatic Creatures

Look deep into nature, and then you will understand everything better.

Albert Einstein

Water covers over two-thirds of the surface area of the Earth. Nearly all of that is saltwater, but salt or fresh, the water is home to an amazing array of creatures that are as expert at moving through water as flying animals are at moving through air. These aquatic creatures and marine mammals interact with hydrodynamic forces on a constant basis. An examination of the effect of hydrodynamic forces on these animals provides an understanding of fluid dynamic principles in action.

For a fish or other animal to move through the water, it must generate sufficient *lift* to equal or exceed the force of *gravity*. It must also be capable of generating sufficient *thrust* to overcome *resistance* and provide forward motion. This can be an exhausting process. As a result, the most economical production of these four forces—*lift*, *gravity*, *thrust*, and *resistance*—is fundamental to an animal's ability to thrive in an aquatic environment.

Lift

Lift is the upward force that counteracts the downward force of gravity. It is an essential force for anything wishing to float above the bottom. Without sufficient lift, a fish will sink. As it sinks, the pressure of the water in the column above the fish will increase, causing the fish to sink at an increasing rate. At some point, the fish will be unable to generate sufficient lift to overcome the pressure of the water column above it. It will not be able to rise again. Fortunately, aquatic animals have several means of creating lift.

The simplest and most energy efficient source of lift is *buoyancy*. It is the force Archimedes discovered while in the baths in 212 BC. Buoyancy is the amount of upward force exerted by a fluid on a body immersed in it (Fig. 3.1). It is equal to the

G. Hagler, *Modeling Ships and Space Craft: The Science and Art of Mastering the Oceans and Sky*, DOI 10.1007/978-1-4614-4596-8_3,
© Springer Science+Business Media, LLC 2013

Fig. 3.1 The buoyancy of an
object is equal to the weight
of the fluid displaced by that
object

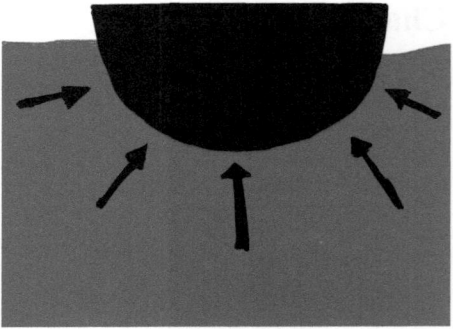

weight of the fluid displaced by the object. When an object's mass equals the mass
of the fluid it displaces, an object is said to be neutrally buoyant. As a result, in calm
conditions, it will remain at a uniform level. If an object is denser than the fluid supporting it, the object will sink. When an object is sinking, it is said to be negatively
buoyant. If rising, it is said to be positively buoyant. Buoyancy in aquatic animals
comes from a variety of sources. Most require an action on the part of the animal,
but for marine mammals, blubber is one source that does not.

Blubber

Blubber acts as a passive mechanism for buoyancy in marine mammals. This dense
vascularized layer of fat beneath the skin is one of their most widely recognized and
universal characteristics.[1] The buoyancy resulting from the presence of blubber
requires no energy expenditure on the part of the mammal. "In most marine mammals (except the sea otter), buoyancy will be determined primarily by the ratio of its
adipose tissue to lean body tissue and body mass. Fat-filled adipose tissue is less
dense than seawater, whereas lean tissue is more dense. Thus, the degree to which
marine mammals store blubber will affect their buoyancy and thus the energy
expended in moving or maintaining position in water."[2]

   The ratio of blubber to lean body tissue and body mass determines the buoyancy
of a marine mammal because the density of the animal for its weight will be far less
with a greater store of blubber than if the animal were predominantly made of tissue
and muscle. The *right whale* has an impressive amount of blubber. Often their blubber layer and the girth exceed 60% of the total body length. Because of this, the
right whale floats after it is dead. In fact, some say that the name, right whale, is
based on the fact that this was the "right" whale to catch because of its slow speed
and the ease of bringing it home after the kill. Others claim it is the "right" whale
because "right" means "true" as in meeting the characteristics of a whale. Whatever
the origin of the name, the right whale is comparatively buoyant.[3]

The large amount of blubber in a right whale and the resulting buoyancy has a downside during a dive and an upside during an ascent. Because of its high degree of buoyancy, "the right whale displays some of its most powerful fluke strokes at the beginning of descent as it counteracts large positive buoyant forces at the start of a dive (Nowacek et al. 2001). The advantage of this positive buoyancy subsequently occurs during the ascent, when the animals are able to glide to the surface and reduce the number of energetically costly strokes."[4]

Whales are not the only marine mammals to benefit from the presence of blubber. Seals, sea lions, dolphins, and porpoises also depend upon fat layers to increase buoyancy. The amount of blubber may change with the location and activities of these animals over the course of the year, but blubber plays a vital role in their insulation and buoyancy control. The presence of blubber also aids in the *streamlining* of marine mammal silhouettes to minimize the resistance encountered as it makes its way through the water.

A source of insulation, blubber helps limit heat loss through the skin. Marine mammals are in constant contact with the water of their marine environment. "As a whole-body envelope of insulation, blubber is central to thermoregulation in marine mammals. Marine mammals, like all mammals, are homeothermic endotherms and hence need maintain a stable body core temperature of about 37°C in cooler (usually <25°C) and often much colder (−1 to 5°C) fluid environments. Additionally, heat is always lost far more rapidly to water than to air because the thermal conductivity of water is 25 times that of air."[5]

Swim Bladder

Many bony fish—those having a skeleton made of bone rather than one made of cartilage—have a *swim bladder*. This gas-filled sac provides buoyancy and helps to keep the fish afloat by keeping it in a neutrally buoyant state. The swim bladder is located in the dorsal portion of the fish and expands or contracts with the pressure exerted upon it in conjunction with a complex process of adjustments to the gas pressure through the use of the gas gland. In less developed types of bony fish, the fish fills or empties the swim bladder by gulping air at the surface. In no case does the gas pass directly through the wall of the swim bladder. Whatever the mechanism for adding or subtracting from the volume of gas in the bladder, it is a source of stability because of its position and horizontal orientation. It supplies lift akin to the manner in which a helium balloon supplies lift.

The wrasse is one type of bony fish that counts on a swim bladder to maintain buoyancy. There are over 600 species of this small, brightly colored marine fish that feed on a wide variety of small invertebrates. The wrasse buries itself in the sand at night, or when frightened. The ability to reduce buoyancy when desired is essential to this behavior. Without it, the wrasse could not rest in the sand on the bottom. The swim bladder also makes it possible for the wrasse to rise and dive in the water, swimming at several different levels while generating sufficient lift to offset the forces of gravity pushing down through the water column above it.

The trout also has a swim bladder to maintain neutral buoyancy. This freshwater fish lives in cool, clear streams and lakes, feeding on other fish and invertebrates. In the trout, the swim bladder is connected to the esophagus. The trout gulps or expels air to maintain the level of gas in the swim bladder. Goldfish and betta are two freshwater fish with swim bladders that often occupy aquariums. Diseases of the swim bladder are not uncommon in these types of fish. When a disease of the swim bladder occurs, the fish will have difficulty regulating its buoyancy, swimming at the top or drifting to the bottom. It may also swim upside down, swim with its head pointing to the bottom of the tank, or swim tilted to one side.

When the swim bladder is functioning properly and the gas capacity is at about 5–7% of a fish's total body volume, the fish will be nearly weightless. Without a swim bladder, a fish would have to swim constantly to generate lift, or live or rest on the ocean floor because a fish is heavier than the water it lives in. Freshwater has a density of 1.0, saltwater has a density of 1.025, and a fish has a density of 1.076 on average. Also, bone is nearly twice as dense as cartilage—1.1 vs. 2.0.

The swim bladder fills an essential role in bony fish since it is not always practical to rest on the bottom, especially in deep water, and swimming constantly is expensive in terms of energy. An associated consideration is that using the fins to generate lift means a fish can't be using them to provide forward thrust. With a swim bladder, neutral buoyancy is ensured and bony fish are free to use their fins for forward motion. There are several ways marine mammals have dealt with this problem. One is to increase body size, which decreases the surface-to-volume ratio and thus provides less surface area per unit volume over which to lose heat. Even the smallest marine mammals are considered large mammals, being one to two orders of magnitude larger than small terrestrial mammals such as rodents and insectivores. Additionally, and perhaps more importantly, large body size generally allows for thicker insulation (be it fur or blubber), which further decreases heat conductance."[6]

"The effectiveness of blubber as an insulative layer depends on its thickness, lipid content, and lipid composition."[7] Because of this it will be more or less effective for different marine mammals depending upon the factors at play, but in general, it will act as an adequate insulator, protecting the marine mammal from heat loss and the need to expend energy on heat production.

Air in the Lungs

For sea turtles, the air in their lungs is the primary source of buoyancy. They are not alone in achieving buoyancy via the air in their lungs. Many marine mammals, especially large marine mammals such as whales, also have large amounts of air in their lungs. This can make a significant difference in the amount of lift generated without the expenditure of a significant amount of energy that could otherwise be used for forward motion.

The lungs of terrestrial and marine mammals work in a similar fashion—up to a point. The difference is that marine mammals make more efficient use of their lung capacity. "The tidal volume (the amount of air breathed in or out during normal respiration) is a larger proportion of the total lung capacity (TLC) in marine mammals than it is of terrestrial mammals. In a typical terrestrial mammal the volume of air inhaled and exhaled in one breath is in the range of 10–15% of TLC. In marine mammals, tidal volume is typically greater than 75% of TLC. The maximum tidal volume or vital capacity (VC) in terrestrial mammals is not more than 75% of TLC, whereas in marine mammals the VC can exceed 90% of TLC."[8] By utilizing a substantially greater amount of TLC, the air in the lungs is another source of buoyancy—and therefore lift—for the marine mammal.

The lungs in marine mammals also differ from the lungs of terrestrial mammals in another way. During a dive, the oxygen in a marine mammal's lungs is not lost, even as the mammal descends and the hydrostatic pressure on the lungs increases. It is at this point that the lungs in marine mammals are protected by a significant adaptation. "Marine mammal lungs contain more elastic tissue than those of terrestrial mammals (Kooyman and Sinnett 1976). The ribs contain more cartilage and are thus more compliant than those of terrestrial mammals. The lung is also more compliant. Marine mammal lungs can collapse and reinflate repeatedly, whereas in terrestrial mammals, lung collapse is a serious situation that requires intervention to reinflate."[9]

The air is not lost when the lungs deflate. Marine mammals compress the air into a reinforced air passageway such as the bronchi or trachea during the dive. The alveoli compress so no nitrogen can go into the blood. As a result, the animals do not get decompression sickness, also known as the bends. On ascent, as the pressure from the water column lessens, the air in the non-respiratory passageways expands and alveoli reinflate.[10]

Another variation from terrestrial mammal lung function is the manner in which marine mammals safeguard themselves from the introduction of large quantities of nitrogen into the blood. "Every inspiration that fills the lungs with air brings in four times as much nitrogen as oxygen. Because nitrogen is neither bound to a carrier in the blood nor metabolized in the tissues, the partial pressure of nitrogen will equilibrate with that in the lungs. If gas exchange is allowed to take place during a dive, the resulting higher partial pressure of nitrogen in the blood and tissues will result in the formation of nitrogen gas bubbles when the external pressure is reduced as the animal comes to the surface. Thus deep-diving marine mammals limit the exchange of gas from lungs to blood during dives."[11]

Air in the lungs and buoyancy provide the majority of the lift required by marine mammals. The swim bladder provides buoyancy and lift for most of the bony fish. For fish without these physical attributes, there are other sources of buoyancy.

Oil in the Liver

Sharks do not have a swim bladder. They are also lacking blubber and air in the lungs. Yet sharks must generate lift if they are to remain neutrally buoyant. One passive way sharks can increase buoyancy is by keeping large quantities of a low-density oil known as squalene in their enlarged livers. This provides lift. In fact, the basking shark has such a large liver with such a large quantity of squalene that it floats when it is dead.[12]

For several deep-sea sharks it was discovered that "the hydrocarbon squalene, which is not a convenient material to have as a metabolic reserve but which, with its low specific gravity (0.86), is particularly suited to give lift.... It is calculated that because of this unusual oil such fish not only obtain the lift needed for neutral buoyancy more economically in terms of the weight of oil required but also in terms of the metabolic energy which has to be used to provide the oil-store responsible for buoyancy."[13]

Oil in the liver is not the main factor in lift for most sharks. The key to lift is movement, along with the effects of a cartilaginous skeleton.

Cartilaginous Skeleton

Most fish have a full skeletal structure consisting of calcified bone. (Humans also have a skeletal structure made of calcified bone.) These fish fall into Class Osteichthyes. Their bones are relatively heavy and inflexible. They have a rib cage, a skull with about 63 bones, and a protective plate covering their gills.

Sharks, rays, and chimeras belong to Class Chondrichthyes. This class of fish is also known as cartilaginous fish because their skeletons are made of cartilage. Cartilage is flexible and lightweight. Cartilaginous fish have no rib cage. They have only ten cartilaginous elements in their skulls and their gill slits are exposed and visible.

There is an advantage to a skeletal system comprised of cartilage rather than bone. It reduces the weight of the shark. This reduction in weight versus the size of the fish reduces the density of the fish. This results in increased buoyancy, or lift, without an expenditure of energy on the part of the cartilaginous fish.

Movement

Movement is yet another mechanism for providing lift in a variety of species. For cartilaginous fish such as the shark, it is the primary source of lift. Without a swim bladder to give it an assist, lungs to fill with air, or blubber to increase its buoyancy, a shark must move constantly. Its streamlined shape keeps resistance to a minimum.

Its skeletal structure made of lightweight cartilage rather than calcified bone reduces its density. None of this is enough to keep a shark neutrally buoyant in the absence of forward movement. Simply put, a shark will begin to sink if it stops moving.

Perpetual movement is exhausting. The shark has evolved into an efficient swimmer. The tail acts as a propeller. The shark swings it back and forth to provide thrust—forward motion. As the shark passes through the water, the water flows over the shark and its fins. The movement of water around the fins creates lift as the fins act as airfoils. This is because the water passing over and under the fins is not moving at the same speed. The water moving over the upper surface of the fin has a longer distance to cover because of the *camber*, or curve, of the upper surface of the fin. As a result, the water has to move more swiftly to keep up with the water passing under the flat bottom surface of the fin. The faster movement of the water over the top of the fin results in decreased pressure above the fin as described by the Bernoulli Principle. This area of low pressure makes it possible for the relatively higher pressure beneath the fin to exert an upward force on the fin, resulting in increased lift.

To increase the amount of lift generated by forward movement, a shark can change the *angle of attack* of its fins. This means it can alter the angle at which the moving water meets the fin. The ability to change the angle of its two sets of paired fins on the sides of the body also allows the shark to position the fins so that there will be greater pressure above the fin than below it when downward motion is desired.

The shark has been in existence in some form or another for over 400 million years. It is understandable that over the course of that time, those sharks that were able to meet their needs with the least amount of energy expended or injury incurred would be the sharks that would survive and go on to breed. As a result of this activity, today's sharks are nearly flawless killing machines. They move through the water in a seemingly effortless manner that nevertheless makes excellent use of the hydrodynamic forces available to it.

Gravity

Water Column

Gravity is the force of attraction between two objects. In the case of aquatic creatures, it is the attraction between that creature and the Earth. The strength of the attraction depends upon the mass of the objects and the distance between the objects. For aquatic animals, the force of gravity is expressed through the *water column* above the aquatic animal. The water column is the imaginary column of water extending from the surface to the bottom sediments of the body of water under study. The pressure in the water column is *hydrostatic pressure*—the pressure exerted by a fluid at equilibrium due to the force of *gravity*.

At some point, depending upon the buoyancy of the individual animal, the pressure experienced from the water column will accelerate at an increasing rate as the buoyancy of the animal decreases. If the aquatic animal is not able to attain and maintain neutral or positive buoyancy in the face of this increasing pressure, it will continue to sink. Aquatic creatures depend upon their ability to generate lift through a variety of means. Their ability to generate lift is essential to maintaining buoyancy in sufficient quantities to offset the forces of gravity and downward pressure from the water column.

Gravity and hydrostatic pressure from the water column has a significant effect during the ascent for mammalian divers. "The beginning of the ascent represents the period of greatest swimming effort for mammalian divers. During this period, many species of pinniped and cetacean use sequential, large amplitude strokes to begin moving upward. As the ascent continues, the physical forces impacting the diver are once again altered as they move through the water column. Hydrostatic pressure decreases on ascent. Consequently, the lungs are able to reinflate and the buoyancy of the marine mammal increases. Swimming behavior reflects these changes with the result that the continuous stroking phase is followed by a stroke and glide mode of swimming, and finally a brief glide to the water surface."[14]

When lift through whatever means is once again sufficient to offset gravity, the animal or object will neither rise nor fall.

Thrust

Thrust is the force that provides forward motion. The force must be greater than the force of resistance. In aquatic animals, thrust is most often generated by the tail. It may be provided by swinging the tail back and forth like a shark. It may be provided by undulating up and down like a dolphin. Whatever the means, there must be sufficient thrust to overcome the forces of resistance, and that thrust must be generated without exhausting the energy supply of the animal seeking forward motion.

Some animals, such as the dolphin, catch a ride by surfing the bow wave of ships. Taking advantage of the moving water eliminates the need for active swimming on the part of the dolphins and results in a significant energy savings. Riding the bow wave and any other activity that provides forward momentum provides thrust. The difficulty is in generating sufficient thrust to counteract the forces of resistance.

Sea turtles generate their thrust with motions that make it appear they are flying through the water. Their graceful motions are silent as they proceed, buoyancy controlled by the air in their lungs and their direction controlled by their rear appendages. These large, gentle creatures are free to use their front appendages for the production of thrust. They use underwater currents to their advantage to expend less energy on migratory travels. The rounded surfaces of their flippers assist in the generation of lift as thrust is generated. These highly efficient animals produce far more torque in the downstroke than in the upstroke. They use their two front semi-rigid, broad and flat flippers to propel their rigid bodies. While the forelimbs are used to produce thrust, the hind limbs are used as rudders. These strokes employ a lift-based mechanism of generated thrust, as confirmed by the angle of attack measurements…[15]

Fig. 3.2 Resistance acts opposite the forward motion of a solid object moving through a fluid

For nearly buoyant animals the results… show that employing such a one-sided power stroke does not constitute a hydrodynamic handicap; the thrust and efficiency remain high and the total average lift force coefficient stays relatively small.[16] For the sea turtle this results in lift generation through the same motion that generates thrust.

Resistance

Resistance is the force that acts in opposition to the forward motion of a solid object making its way through a liquid (Fig. 3.2). In his seminal work, "The Speed and Power of Ships," Rear Admiral David W. Taylor identified four forces that make up resistance. Admiral Taylor wrote, "There are several kinds of resistance and usually all are present in the case of every ship. They will be enumerated here...

§ 32. *Skin or Frictional Resistance.* In the first place, water is not frictionless. Its motion past the surface of the ship involves a certain amount of frictional drag, the resistance of the surface involving an equal and opposite pull upon the water.

This kind of resistance is conveniently denoted by the term skin resistance. It is nearly always the most important factor of the total resistance.

§ 33. *Eddy Resistance.* While skin resistance is accompanied by eddies or vortices in the water near the ship's surface, the expression eddy resistance is used for a different kind of resistance. The motion through the water of a blunt or square stern post or of a short and thick strut arm, etc., is accompanied by much resistance and the tailing aft of a mass of eddying confused water. Such resistance is designated eddy resistance. With proper design it is in most cases but a minor factor of the total resistance.

§ 34. *Wave Resistance.* A far more important factor, which though usually second to the skin resistance is in some cases the largest single factor in the total resistance, is the resistance due to the waves created by the motion of the ship. It is called for brevity the wave resistance.

We have seen that the motion of a ship through the water is accompanied by the production of surface waves. These absorb energy in their production and propagation, and this energy is communicated to them from the ship, being derived from the wave resistance.

§ 35. *Air Resistance.* Finally, we have the air resistance, which is, as its name implies, the resistance which the air offers to the motion of the ship through it. The air resistance is seldom large. It is, however, by no means always negligible and of late years it has been found necessary to make allowance for it if we wish to make accurate analyses of trails."[17]

Taylor's work dealt with the resistance encountered by vessels moving through the water, but it has direct application to aquatic animals and marine mammals. When Admiral Taylor said that water is not frictionless he was referring to the *viscosity* of water. Viscosity is a measure of the internal friction of a fluid. It's the reason molasses flows more slowly than water. It's the reason cold molasses flows more slowly than warm molasses. The viscosity of air is less than the viscosity of water, which is one of the reasons it is generally easier to move through air than through water.

Viscosity is subject to change due to changes in temperature and pressure in *Newtonian fluids* such as water. This is because the relationship between the *shear stress*—a stress that is parallel to the fluid flow—and *shear rate*—a measure of the ease with which the parallel internal surfaces of a fluid slide past each other—is linear. There is a definable coefficient of viscosity and that coefficient is constant. Any other forces acting on a Newtonian fluid will have no effect. Because of this, the behavior of a Newtonian fluid is predictable.

A *Non-Newtonian fluid* is one that does not behave like a Newtonian fluid. It is subject to shear stress from factors other than temperature and pressure. Because of this, its behavior is not predictable and it's not possible to calculate a coefficient of viscosity. Even at that, not all Non-Newtonian fluids behave identically to forces other than temperature and pressure. Non-Newtonian fluids that become less viscous with increasing shear stress are said to be "time-independent shear thinning" substances. Ketchup, syrup, and molasses are examples. Cornstarch and water mixtures are said to be "time-independent shear thickening" substances. Jump on or punch this solution and it becomes a solid. Reduce the pressure and it revert backs to a liquid.

The *wetted surface* of a solid moving through the water is the portion that is in the water. For a ship, this is the portion of the ship below the waterline. For a fish that spends its time beneath the water, it is the total body of the fish. With a ship, the wetted surface may vary greatly. It may sit higher in the water when it is not carrying cargo, with a smaller wetted surface as a result, than it does when it is full and sitting deeper in the water. The amount of surface area comprising the wetted surface is significant because it acts as the wall that forms the *boundary layer*.

Ludwig Prandtl was the one to define the boundary layer in 1904. It is made up of the area that includes the side of the object immersed in the fluid or with the fluid running through it. It can be the interior of a pipe with water, air, or gasses flowing through it. It can be the skin of a fish immersed in water. The fluid that is closest to the interior of the pipe or the skin of the fish is not moving. This is because of high levels of friction where the water and the skin meet. The water that is adjacent to the water that is not moving—the water parallel to it in the streamline—moves a bit, but still slowly. The water adjacent to that water behaves in a similar manner. This behavior continues until the effect of the force of friction is diminished sufficiently for the moving water to flow at a normal speed. The area where this slowing occurs is the area of frictional resistance. The more the water slows in the boundary layer, the more significant the resistance.

However, no matter how great the resistance at the boundary layer, some portion of the fluid will still flow. Fluids are said to flow *continuously*, without any gaps in the flow. The molecules in a fluid flow in close relation to one another in a *continuum*. Because of this, a fluid will flow or run smoothly with unbroken continuity in what has been termed a *streamline*. In the streamline, a continuous series of particles follow each other in an orderly fashion in parallel with other streamlines. For the contemporary imagination, it may be helpful to think of water having layers and flowing like parallel sets of bytes of information with eight bits traveling alongside one another and each bit following the one before it. In an *ideal fluid* (which does not actually exist) each streamline would maintain its position unchanged in a steady current. In a *real fluid* there will be events that interrupt the steady flow of the fluid.

An understanding of viscosity and boundary layers is significant to an understanding of resistance. Whether the source of frictional resistance is from the skin or wetted surface of a solid object such as a fish, ripples created by a solid object as it moves through the water, the waves created by a fish when it moves near the surface, or the air when a fish leaps out of the water, it all impedes forward motion.

Overcoming Resistance

Overcoming resistance is a primary consideration for anything seeking movement through a fluid. The reward for a reduction in resistance is a decrease in the energy required for forward movement. Resistance can be overcome through the application of force, but that is a highly inefficient and exhausting process. There are several other ways to reduce resistance. Some methods are better at meeting the needs of one type of aquatic animal than another.

High Swimming Speeds

High swimming speeds attained by many marine mammals have focused attention toward the possibility of specialized drag reduction mechanisms. In Gray's Paradox hydrodynamic estimates of dolphin power output at high speeds were inferred to be greater than the power that could be developed for the mass of muscle available for swimming (Gray 1936). Resolution of the paradox was believed only possible if the drag was reduced by maintaining laminar flow within the boundary layer, despite a high swimming speed dictating a turbulent boundary flow with increased viscous drag. To date there has been no conclusive proof of this phenomenon. Investigations have included examinations of the mechanisms of compliant skin dampening, secretions, skin cell sloughing, infusion of long-chain polymers into the boundary layer, boundary layer heating, skin folds, and boundary layer acceleration (Fish and Hui 1991).[18]

It's possible special drag reduction mechanisms are unnecessary. Gray's Paradox, is reconciled when one considers that the calculation of power output were based on burst swimming (10 m/s for 7 s) and muscle power output was underestimated, because it was based on sustained performance of dogs and humans.[19] In short, Gray's Paradox may not have been a paradox at all. It may simply have been the result of an underestimation of the dolphin's capacity for sustained, high-speed forward motion.

Streamlining

Streamlining is not the same as the *streamline* concept of fluid flows put forth by William John Macquorn Rankine at the end of the nineteenth century. Rankine's theory of streamlines described the way water flows with one particle following another, in parallel with other particles unless it is disturbed by an outside force. Streamlining in marine mammals is instead a component of hydrodynamic efficiency having to do with their body shape and the flow of water around that shape. Streamlining in marine mammals is of vital importance for two reasons. First, marine animals are surrounded by water at all times. It's most often the case that

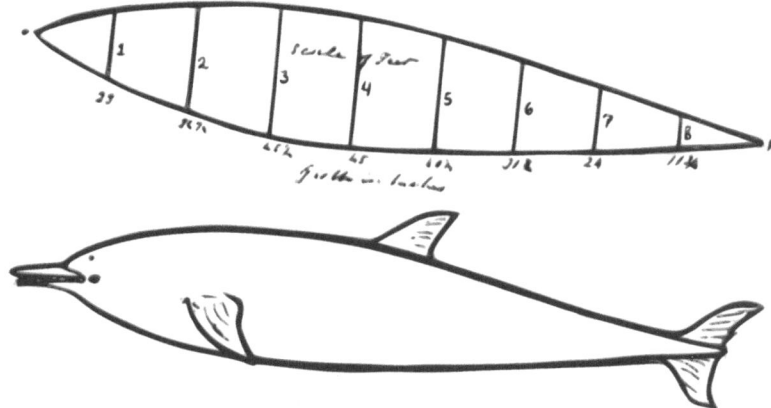

Fig. 3.3 Cayley's dolphin. Sir George Cayley first explored the dolphin as a solid with a shape of least resistance in the 1800s

their entire bodies are below the waterline. Because of this, it is imperative they are as hydrodynamically efficient as possible. The second reason is that, "the single most effective way to reduce both drag and the power required for forward motion through a fluid is to have a smooth streamlined shape. Although all marine mammals tend to be somewhat streamlined in body shape as defined by their musculoskeletal system, blubber provides their form with a smooth sculpted contour."[20]

Overcoming the effects of pressure or form drag and the viscous or skin friction drag is especially important to marine mammals because so much of the surface area of their bodies is subject to the impact of the boundary layer. "Water particles adhere to the body surface within a thin layer of water adjacent to the body, called the boundary layer. Friction within the boundary layer and between the boundary layer and the body create a force in the drag direction. The magnitude of the viscous drag will depend on the wetted surface area of the body and the flow condition within the boundary layer."[21]

For marine mammals, streamlining is a matter of body shaping. "The streamlined profile of these structure has a fusiform design resembling an elongate teardrop with a rounded leading edge extending to a maximum thickness and a slowly tapering tail. This shape was first investigated in the dolphin by Sir George Cayley (circa 1800) as a solid of least resistance design… (Fig. 3.3). This fusiform shape is sculpted by the distribution of blubber and/or fur covering the body."[22] This is the same Sir George Cayley who was the first to identify the four forces of flight. Not incidentally, since air and water are both fluids, the forces Cayley identified were lift, gravity, thrust, and resistance.

Bony fish may have a body shape that is hydrodynamically efficient but this is not accomplished through the use of blubber. Cartilaginous fish are also hydrodynamically efficient. This is also accomplished without the presence of blubber.

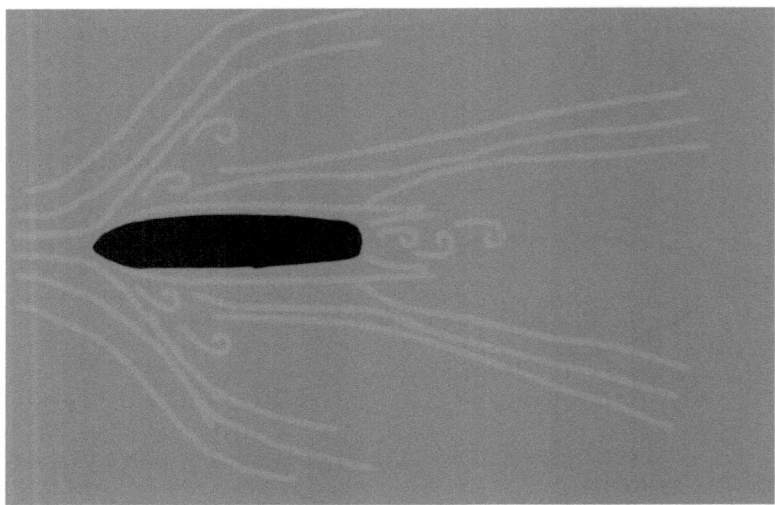

Fig. 3.4 The Reynold's Number assists in the identification of the point at which a laminar flow will change to a turbulent flow

Objects moving through a fluid initially generate a laminar boundary layer at the bow that flows along the wetted surface increasing its thickness as it progresses down the side. At some distance the flow changes to turbulent flow according to the *Reynolds number* and surface conditions. In the end, it separates due to the pressure gradient and produces a wake that modifies the pressure field around the body and produces drag (Fig. 3.4). Because of this, a design goal is to have sufficient length in the laminar and turbulent flows before the separation point where the rear turbulence is experienced and further impacts performance. With a streamlined body, the position of maximum thickness is called the shoulder. The shoulder position is important because it is here that the flow turns from laminar to turbulent. It is also here that boundary layer separation occurs. The shoulder position for dolphins is 34–45% of the body length from the back.[23] With the shoulders located this far back, there is less resistance and increased energy efficiency for the dolphins.

"Experiments on flow visualization using a fluorescent dye applied to a dolphin's melon [the rounded region of the forehead] showed the flow to be laminar over the anterior 32% of the dolphin. Transition began before the dorsal fin with turbulence aft of the fin. Separation of the boundary flow occurred smoothly near the base of the flukes. Flow visualization using bioluminescence within the boundary layer of dolphins and seals similarly indicated a lack of separation from the body surface (Fish 2006). Flow separation is restricted to the tips of the flukes, flippers, and dorsal fin. The flow separation has been observed as bioluminescent 'contrails.'

The contrails are vortices generated at the tips of the appendages. A tip vortex is generated from pressure differences along the two surfaces of the appendage. The pressure difference produces a lift force similar to the lift produced by airplane wings."[24] This lift is another benefit of streamlining for marine mammals.

Skin Surface and Dermal Denticles

Another method for reducing resistance is for the object moving through it to have as few flaws as possible. This will reduce the opportunity for friction by presenting a smooth surface to the hydrodynamic forces in effect. Smoother skin will result in a smoother flow, which will result in reduced resistance. "The naked skin of cetaceans is regarded as a means to maintain a smooth flow with an attached boundary layer over the surface of the body. In addition, the cells of epidermis are produced rapidly, which promotes a high rate of skin sloughing. This increased skin sloughing deters organisms, such as barnacles, from attaching to the skin and thus minimizes drag."[25]

It isn't just the smooth skin of marine mammals that can make a difference. The arrangement of scales on bony fish also reduces drag. The placement of the gills, as well as the existence of any scales covering the gills, also assists in reducing resistance. Anything that facilitates the smooth flow of water over the body from end to end is something that will increase fluid dynamic efficiency and reduce the energy expended to overcome the forces of resistance.

Sharks do not have smooth skin. They have dermal denticles covering their body surface. Because these cartilaginous fish are covered with these placoid scales, the skin is protected from parasites and damage. Hydrodynamic efficiency is also increased through the creation of small vortices that reduce drag as water flows across the surface of the denticles. An ancillary benefit of the action of the dermal denticles is that they render sharks silent as they move through the water, contributing to their success as hunters.

Porpoising and Free-Riding Behaviors

It is more decidedly difficult to move through water than air for a variety of reasons. The water is not only more viscous; it is also denser. Adding to this is the fact that a swimmer moving through the water increases its body drag exponentially.[26] Combine the effects of density and viscosity to this increasing body drag and it's clear that there is an advantage to be gained by having some of the forward movement that takes place, take place in the air rather than in the water.

To accomplish this, many animals leap at least partly into the air in a move known as *porpoising* as they move forward at a high rate of speed. Penguins, dolphins, and whales all do this. It is effective only when the animals are moving quickly because there is energy expended to rise and clear the surface. This move only saves energy if the energy expended to reach the surface and exit into the air is less than the energy gain from covering a portion of the total distance to be traveled outside of the water.

As well, "many dolphins utilize free-riding behaviors to reduce the energy cost of swimming (Williams et al. 1992). In this behavior, the dolphin takes advantage of the pressure field generated by another body and moves along with little or no energetic input. Dolphins have been observed to ride the pressure waves of ships and large whales. By situating itself on the bow wave, the small cetacean can be pushed along or surf down the front slope of the wave… Even large whales my reduce swimming effort by using the energy of large oceanic waves."[27]

Schooling

For smaller fish, traveling in schools is one method of decreasing resistance. The principle in action with this behavior is similar to the drafting behavior of Canada geese in their v-formation or cyclists riding in single file. The fish behind or adjacent to other fish are required to expend less energy for forward movement because they take advantage of a portion of the lift generated by the fish that precede them. This is true even in the case of fish with a swim bladder, since lift is generated not only from the swim bladder, but through the Bernoulli effect as water flows over the fins of the fish as they are in motion. Studies about the role of swimming in schools are inconclusive but it seems likely that fish are striving to expend as little energy as possible on the basics of their mobility.

Conclusion

Once we shift our perception of the oceans or manage to imagine ourselves standing at the bottom of them as we do at the bottom of our ocean of air, it's simple to envision the four forces of flight in action. After all, these forces apply to fluid flows and water is certainly a fluid. The creatures thriving above us in the water column fall into strata mimicking that of the animals we observe moving through the air. There are bottom dwellers like crabs, clams, and lobsters to take the place of land animals such as mice, worms, and squirrels. There are fish and sea turtles that maintain a position nearer the surface like our birds. There are those fish nearer still to the surface that take the place of our airplanes. It's even possible to liken those aquatic animals leaping out of the water to craft that leave the air and make their way into the vacuum of space.

All these aquatic creatures need to generate lift to overcome the forces of gravity. They need to overcome resistance with sufficient thrust to achieve forward momentum. They need mechanisms for controlling their movement while reducing the amount of energy required to produce that movement. All in all, it is not so different to be moving through the water than moving through the air. The principles of fluid dynamics anticipate that similarity and provide us with a basis for understanding, appreciating, and working in an aquatic environment.

Notes

1. Perrin, W. F., B. G. Würsig, et al. (2009). *Encyclopedia of marine mammals*. San Diego, Academic Press.
2. Ibid.
3. Ibid.
4. Ibid.
5. Ibid.
6. Ramel, G. "Fish Anatomy: The Swim Bladder." *Earthlife Web*. from http://www.earthlife. net/fish/bladder.html.
7. Perrin, W. F., B. G. Würsig, et al. (2009). *Encyclopedia of marine mammals*. San Diego, Academic Press.
8. Ibid.
9. Ibid.
10. Ibid.
11. Perrin, W. F., B. G. Würsig, et al. (2009). *Encyclopedia of marine mammals*. San Diego, Academic Press.
12. Fish, F. (2012). Buoyancy in Marine Mammals and Reptiles. G. Hagler.
13. Corner, E. D. S. D., E.J.; Forster, G.R. (1969). "On the Buoyancy of Some Deep-Sea Sharks." *Proceedings of the Royal Society of London Biological Sciences* **171**(1025): 415–429.
14. Perrin, W. F., B. G. Würsig, et al. (2009). *Encyclopedia of marine mammals*. San Diego, Academic Press.
15. Licht, S. C. W., M. S.; Hover, F. S.; and Triantafyllou, M. S. (2009). "In-line motion causes high thrust and efficiendy in flapping foils that use power downstroke." *The Journal of Experimental Biology* **213**: 63–71.
16. Ibid.
17. Taylor, D. W., United States. Maritime Commission. [from old catalog], et al. (1943). *The speed and power of ships; a manual of marine propulsion*. Washington,, U. S. Govt. Print. Off.
18. Perrin, W. F., B. G. Würsig, et al. (2009). *Encyclopedia of marine mammals*. San Diego, Academic Press.
19. Ibid.
20. Ibid.
21. Ibid.
22. Ibid.
23. Ibid.
24. Ibid.
25. Ibid.
26. Ibid.
27. Ibid.

Part II
Evolution of Theory

Observing fluid dynamic principles in action is one thing. Developing an understanding of what is being observed is another. The following hydrodynamic and aerodynamic theorists advanced the science of their time through careful investigation. Some of their work was known during their own time, much of it was not. Each worked on a problem of interest to him because of his work or simply as a matter of intellectual curiosity. Some individuals investigated phenomenon that falls clearly into the theory of hydrodynamics or aerodynamics. Some individuals investigated phenomenon that contributed to both fields.

The men whose work is described were innovators who developed the methods necessary for the studies they wished to conduct. When there was no established way to communicate their findings, they invented new fields of mathematics and science. The sum total of their work forms the basis for the science of fluids that is in use today.

Chapter 4
Hydrodynamic Theorists

It is called Hydrodynamics when the fluid is water.

It's easy enough to examine the principles of current science and assume it was always obvious to theorists that their work was vital to a greater whole, but this is not the case. A look at the work done by generations of hydrodynamic theorists shows how the labor of each added to the science of the day, ultimately leading to scale model testing in the basins of Froude and Taylor.

It's doubtful Aristotle and Archimedes foresaw a day when their work would inform the science used to design and test ships, aircraft, and spacecraft making their way through the ocean and sky. Yet ships, aircraft, and spacecraft are all subject to the principles of fluids in motion, and the theories forming the foundation of the science behind them began with these early thinkers.

Aristotle Through Hero: Early Theorists

Aristotle (384–322 BC) was one of the greatest philosophers and scientists of all time. A student of Plato and a teacher of Alexander the Great, this ancient Greek's interests were far ranging. Although he had no way of knowing at the time, they included many that touched on the basics of hydrodynamics.

Aristotle began his studies at Plato's Academy when he was 17. At first he concentrated on questions dealing with morality and existence. While in his 30s, he turned his attention to the natural world and the way it functioned. He was exceptionally observant, and his work culminated in the *Historia Animalium* (*Research into Animals*), a work that describes animals ranging from insects to woodpeckers. It includes not only descriptions of the animals, but of their habitats and behaviors as well.

G. Hagler, *Modeling Ships and Space Craft: The Science and Art of Mastering the Oceans and Sky*, DOI 10.1007/978-1-4614-4596-8_4,
© Springer Science+Business Media, LLC 2013

Fig. 4.1 Aristotle's rate of fall. Aristotle incorrectly theorized that an object that was twice as heavy as another with the same shape and size would fall at twice the speed of the lighter object

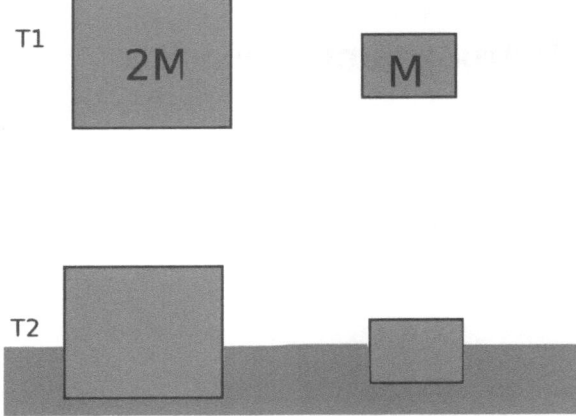

Aristotle also developed a classification system for all known plants and animals. The system was based on their physical characteristics, differentiating between those with and without backbones for example, and was used until Swedish naturalist Carl Linnaeus created a more detailed system in the 1700s.

Not everything Aristotle concluded was correct; however, his reputation was formidable. When he incorrectly asserted that nature did not tolerate "nothing," and, therefore, a vacuum could not exist, it was generally believed that he was correct. He was also incorrect when he posited the idea that if two objects of the same shape and size, one weighing x and the other $2x$, were dropped from the same height at the same time, the heavier object would reach the ground in half the time. That assertion was accepted for nearly 2,000 years, however, until Galileo proved that bodies of all weights fall from the same height in the same time (Fig. 4.1).

Aristotle did make several correct observations about fluids in motion that would be of great importance to the future study of fluids. Among them, his assertion that there is a "heaviness" to bodies and that they tend to fall toward their "natural place" led him to correctly conclude that the lower layers of matter, e.g., water, must be more dense than the upper layers, e.g., the atmosphere.

He correctly put forth the concept of *continuum*, the idea that a fluid completely fills the space it occupies. This assumption about the nature of fluids is a fundamental part of fluid dynamics today because it underlies the view of fluids as a continuous substance that can be tested at any point, rather than as a substance that will show different characteristics depending upon where it is tested. Aristotle also correctly understood that "something" must work on a body in motion to bring it to rest. The "something" is known today as *resistance*. The observations he made about the ease of flow of various liquids is integral to an understanding of what is known today as *viscosity*.

Clearly, these principles are also fundamental to an understanding of the movement of objects through fluids like water and air, and Aristotle further observed the behavior of objects moving through air. He described the way these objects become

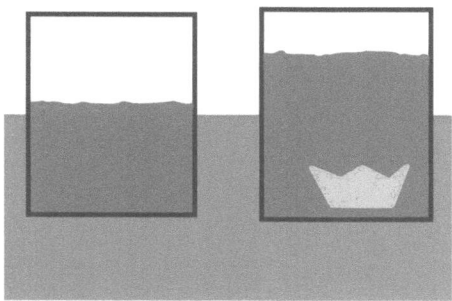

hot and sometimes even melt, leading to an understanding of kinetic heating. He was the first to put forth a law of inertia, as well, when he wrote about an object staying in motion until it encountered a force bringing it to rest.

Another ancient Greek, Archimedes (287–212 BC), established the principles of plain and solid geometry. He was also the first to find a workable approximation of *pi*—the ratio between the diameter and circumference of a circle. His use of the method of exhaustion to estimate the area under curves was a precursor to the branch of mathematics we know as calculus.

"In mechanics he defined the principle of the lever and is credited with inventing the compound pulley and the hydraulic screw for raising water from a lower to higher level... During the Roman conquest of Sicily in 214 BC Archimedes worked for the state, and several of his mechanical devices were employed in the defence of Syracuse. Among the war machines attributed to him are the catapult and—perhaps legendary—a mirror system for focusing the sun's rays on the invaders' boats and igniting them. After Syracuse was captured, Archimedes was killed by a Roman soldier. It is said that he was so absorbed in his calculations he told his killer not to disturb him."[1]

Archimedes also proved that a sphere's volume is equal to two-third of the volume of the smallest cylinder in which that sphere will fit. This is because there is a 2 to 3 ratio between the sphere and the cylinder. He was so proud of this work that he asked that a figure of the sphere and cylinder be marked on his gravestone.

In the most famous story about Archimedes, it is said that one day while in the public baths, he noticed a force pushing up against his body. The more he immersed himself, the greater the force pushing up against him. He also noticed that when he stepped into the bath, the level of water in the bath rose. When he stepped out, the level returned to where it had been. When he realized he could apply what he was observing to a problem he'd been tasked with solving for King Heiro II, Archimedes is said to have run naked through the streets of Syracuse, exclaiming "Eureka!" (I have found it!)

Buoyancy was Archimedes' experience of feeling pushed up by a force in the water. In fact, the force was equal to the weight of the water that was *displaced* by his body. *Displacement* defines the property of a body, when immersed in a fluid, to push the fluid out of the way and occupy the space (Fig. 4.2). It explains what happened when Archimedes stepped into the bath and the water level rose. His observation

Fig. 4.3 Hero's aeolipile.
Hero's aeolipile is believed
to be the first steam-powered
engine

is known today as *Archimedes' Principle*: When an object is immersed in a fluid, it
is buoyed upward by a force equal to the weight of the fluid displaced.

In the anecdote about Archimedes and his bath, King Heiro II suspected that
silver had been substituted for some of the gold in a crown he'd had made. He left
it to Archimedes to prove this. Archimedes used his observation of displacement to
design a simple test to determine whether or not the amount of gold that was sup-
posed to have been used to make the crown had indeed been used.

He submerged an amount of gold equal to what should have been used and measured
the rise in the water. He then submerged the crown and measured the rise in the water.
If the full amount of gold had been used in the crown, the measured rise in the water
would have been equal. It was not, and Archimedes correctly concluded that this was
because another material had been substituted for some of the gold in the crown. With
this exercise, Archimedes not only solved a problem for Heiro II by devising a method
for determining the volume of an irregular shape, but also established the method for
comparing an unknown quantity of material to a known quantity of material.

Archimedes' work with fluids at rest, *hydrostatics*, resulted in a work entitled
"On Floating Bodies." In this work, Archimedes lays out several propositions that
include the definitions of buoyancy and displacement.

Hero of Alexandria (ca. AD 65–125) was another mathematician of ancient
times. His primary area of study was in the field of mechanical devices. His work
also contributed to the foundation of land surveying. He used geometry to solve
problems of length, area, and volume and derived the formula for calculating the
area of a triangle from the length of its three sides. He is best known for his work
with the approximation of square roots.

Hero was quite famous in his own time for his work in mechanics. He invented a
water clock to accurately track the time. He also constructed a catapult that used
compressed air for its energy. He was also one of the first to put the principles of fluid
dynamics to practical use. He did this by constructing an aeolipile. Believed to be the
first steam-powered engine, it consisted of a hollow copper ball filled with water. The
ball had two L-shaped pipes extending from opposite sides and was suspended above
a fire, between two poles that attached to what would be the north and south poles on
a globe (Fig. 4.3). When the water inside the ball began to boil, steam was released

through the L-shaped pipes and the ball spun. Hero's steam-powered ball established steam as a means of propulsion. It would not be for centuries that Robert Fulton would successfully harness steam power to propel his paddlewheel boats.

da Vinci Through Lagrange: Fluid Dynamics

Leonardo da Vinci (1452–1519) was a restless genius, pursuing studies in topics ranging from architecture to the dynamics of water. When he was 14, he was apprenticed to Florentine artist Andrea del Verrocchio. Accepted into the painters' guild at the age of 20, da Vinci continued to work alongside his former master until he opened his own studio in Florence in 1477.

From 1482 to 1500, da Vinci lived in Milan. It was while he was here that he painted *The Last Supper* and undertook work as an engineer and architect for the Duke of Milan. With his illustrations of Luca Pacioli's book on mathematical proportions, Leonardo became interested in geometry. Back in Florence in 1500, da Vinci's interests were largely mathematical and scientific. It was during this time that he studied fluids, flight, and cartography.

"The fame of da Vinci's surviving paintings has meant that he has been regarded primarily as an artist, but the thousands of surviving pages of his notebooks reveal the most eclectic and brilliant of minds. He wrote and drew on subjects including geology, anatomy (which he studied in order to paint the human form more accurately), flight, gravity and optics, often flitting from subject to subject on a single page, and writing in left-handed mirror script. He 'invented' the bicycle, airplane, helicopter, and parachute some 500 years ahead of their time.

If all this work had been published in an intelligible form, da Vinci's place as a pioneering scientist would have been beyond dispute. Yet his true genius was not as a scientist or an artist, but as a combination of the two: an 'artist-engineer.' His painting was scientific, based on a deep understanding of the workings of the human body and the physics of light and shade. His science was expressed through art, and his drawings and diagrams show what he meant, and how he understood the world to work."[2]

Because da Vinci chose to write his notes backward, as well as because he did not order them by topic, his contributions were not well recognized during his lifetime. However, nearly 8,000 pages of his notebooks survive. It is his *Codex Leicester* that includes his observations and theories on the properties of water. Some of his observations resulted in conclusions that were quite simple: "Water does not move unless it descends." Others resulted in findings that were far more elaborate, like the calculation of the volume of water flowing in a river or canal.

This interest in the way water flows in rivers led da Vinci to draw a number of conclusions still in use in hydrodynamics today. The most valuable was his realization that for an *incompressible* fluid like water, the number of pounds per second moving through any part of the system is constant and can be described as AV = constant. The variable A is equal to the cross-sectional area, and V is the

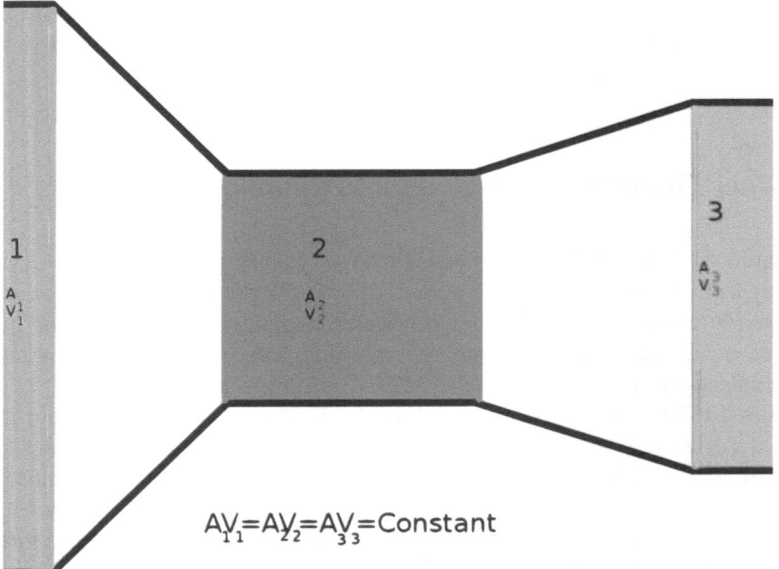

$$A\underset{1}{V_1} = A\underset{2}{V_2} = A\underset{3}{V_3} = \text{Constant}$$

Fig. 4.4 da Vinci AV=constant. The Continuity Equation states that as the area decrease in size, the velocity of the fluid increases, and vice versa

velocity of the fluid at that same location (Fig. 4.4). This *continuity equation* states that when the area becomes smaller, the velocity will increase, and vice versa, to keep the pounds per second moving through the area, moving at a constant rate. The practical application of this principle, as noted in the first chapter, is the way water flows more quickly through a smaller opening in a nozzle than through a larger opening.[3]

Edme Mariotte (1620–1684) and Christian Huygens (1629–1695) contributed the *velocity-squared law* to the science of fluid dynamics. They did not work together but they arrived at the same conclusion during their lifetimes. The *velocity-squared law* states that resistance is not proportional to the velocity; it is the square of the velocity. This realization is deceptively simple. It underlies our intuitive understanding that it is difficult to move a little bit faster when you're already moving very fast, yet easier to go a little bit faster when you're moving slowly. The reason for this is that resistance is not linear and the rate at which fluid resistance increases, increases more quickly than the speed you're moving at (Fig. 4.5). An understanding of this concept is imperative to the design of large ships today.

Robert Boyle (1627–1691) was an Irish chemist best known for his gas laws. He never earned a college degree but he was well educated by private tutors throughout his life. The 14th child of a wealthy and aristocratic family, he was studying Latin and Greek by the age of 8. From the age of 11–16 he traveled through Europe with his tutor. Wealthy enough to pursue his interests, Boyle lived

Sample Velocity Resistance Table, where R is proportional to V^2

Velocity (m/s)	Resistance (N)
0	0
1	1
2	4
3	9
4	16
5	25
6	36

Fig. 4.5 Velocity-squared law. It is difficult to move a little bit faster when you are already moving very quickly

at the University of Oxford from 1656 to 1668. He was not a student but he did participate in meetings of the Invisible College. This was a group of scientists who believed that experimentation was more valuable than the use of logic alone to arrive at conclusions.

It was while at Oxford that Boyle began his experiments on gases. Boyle constructed an air pump that allowed him to produce a vacuum in a sealed container. His experiments with his vacuum chamber and air pump included those on the effect of vacuum on sound, and the necessity of air for respiration and combustion. In 1660 he published his findings in his book, *New Experiments in Physio-Mechanicall, Touching the Spring of the Air and its Effects.*

"At this time even the idea of an experiment was controversial. The established method of 'discovering' something was to argue it out, using the established logical rules Aristotle and others had worked out 2,000 years before. Boyle was more interested in observing nature and drawing his conclusions from what actually happened. He was the first prominent scientist to perform controlled experiments and publish his work with details concerning procedure, apparatus and observations. He began to publish in 1659 and continued to do so for the rest of his life on subjects as diverse as philosophy, medicine and religion.

It is Boyle's Law for which he remains most famous. This states that if the volume of a gas is decreased, the pressure increases proportionally. Understanding that his results could be explained if all gases were made of tiny particles, Boyle tried to construct a universal 'corpuscular theory' of chemistry. He defined the modern idea of an 'element,' as well as introducing the litmus test to tell acids from bases, and introduced many other standard chemical tests.

In 1660, together with 11 others, Boyle formed the Royal Society in London which met to witness experiments and discuss what we would now call scientific topics."[4]

Among Boyle's findings was the fact that air is *compressible*. He also found and published, in 1661, that at a constant temperature, the volume of a gas is inversely proportional to its pressure. This is known today as Boyle's Gas Laws. Boyle also concluded that air is not a continuous substance because the volume of a gas

Newton's First Law

"Every object persists in its state of rest
or uniform motion in a straight line unless
it is compelled to change that state
by forces impressed on it."

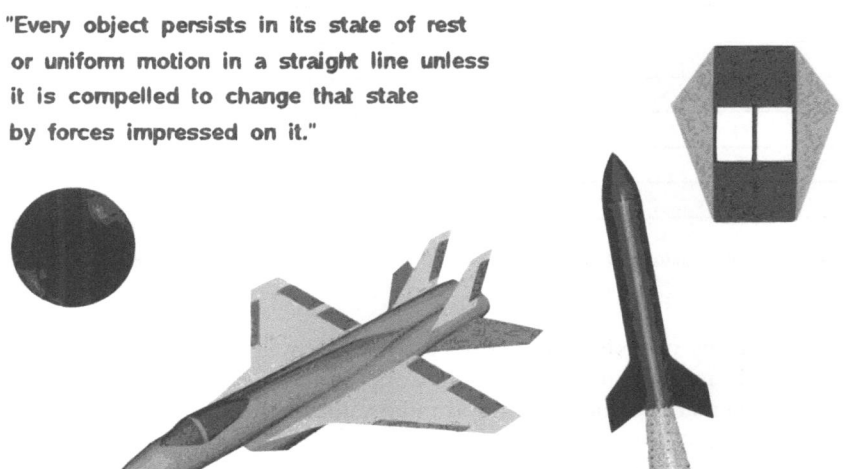

Fig. 4.6 Newton's first law. Newton's First Law of Motion. *Source*: NASA/courtesy of nasaimages.org

decreases as pressure increases. To Boyle this proved that air consists of individual particles separated by empty space.

Boyle also did pioneering work in what is now the field of chemistry. In 1661, he published *The Sceptical Chymist*. This book marked the switch from alchemy to chemistry. Boyle also developed a theory that led to the theory of chemical elements and suggested a method of distinguishing between acids and bases. In 1680 he was elected president of the Royal Society. He refused the position. He was well known as a scientist during his own time.

Sir Isaac Newton (1642–1727) was an English physicist and mathematician. The greatest scientist of his time, his work added three laws of motion to the growing body of knowledge. These laws are the basis for several fundamental concepts of fluid mechanics. *Newton's First Law* is the law of inertia and states that a body in motion moves at a constant velocity covering equal distance in equal time in a straight line until acted upon by another force (Fig. 4.6). *Newton's Second Law* states that the net force exerted on an object is equal to the product of that object's mass times its acceleration and takes place in the direction of the straight line in which the force acts (Fig. 4.7). *Newton's Third Law* states that for every force there is an equal and opposite force. These laws play a fundamental law in many branches of science (Fig. 4.8).

Newton also established the modern study of optics and, "in 1687, with the support of his friend the astronomer Edmond Halley, Newton published his single

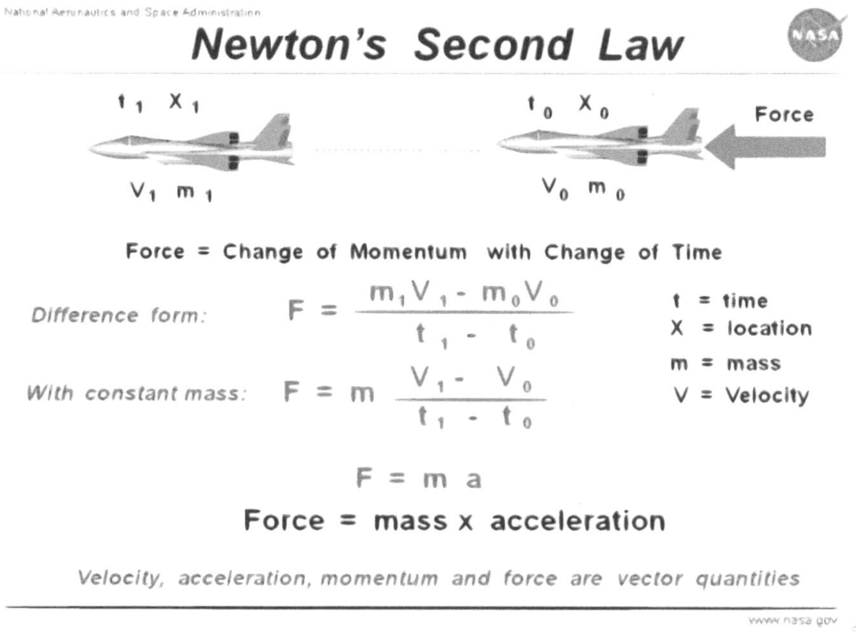

Fig. 4.7 Newton's second law. Newton's Second Law of Motion. *Source*: NASA/courtesy of nasaimages.org

greatest work, the 'Philosophiae Naturalis Principia Mathematica' ('Mathematical Principles of Natural Philosophy'). This showed how a universal force, gravity, applied to all objects in all parts of the universe."[5]

Newton is said to have been a difficult man who was prone to depression. This reputation did not keep him from being elected to a number of prestigious positions, including as president of the Royal Society in 1703. He was also knighted in 1705.[6]

The *Pitot tube*, a simple device still in use today, delivers essential information about fluid flows (Fig. 4.9). Originally its inventor, French hydraulic engineer Henri Pitot (1695–1771), designed the tube to measure the speed of water at a given point in rivers and canals. He introduced his new device in 1735, before the Academy of Sciences. "That invention was motivated by his dissatisfaction with the existing technique for measuring the flow velocity of water, which was to observe the progress of a floating object on the surface of the water. So he devised an instrument consisting of two tubes; one was simply a straight tube, open at one end, that was inserted vertically into the water (to measure static pressure p), and the other was a tube with one end bent at a right angle, with the open end facing directly into the flow (to measure the total pressure p_o). From a bridge over the Seine River in Paris, he used the instrument to measure the flow velocity at different depths within the river... Contemporary theory, based on the experience of some Italian engineers, held that the flow velocity at a given depth in a river was proportional to the mass

Newton's Third Law
Applied to Aerodynamics

Glenn
Research
Center

For every action, there is an equal and opposite re-action.

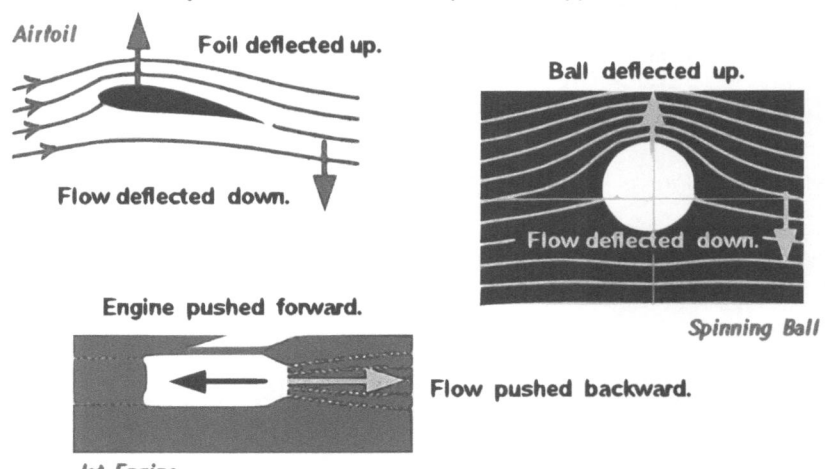

Fig. 4.8 Newton's third law. Newton's Third Law of Motion. *Source*: NASA/courtesy of nasaimages.org

Fig. 4.9 Pitot tube. The pitot tube is used to measure the speed of a moving aircraft

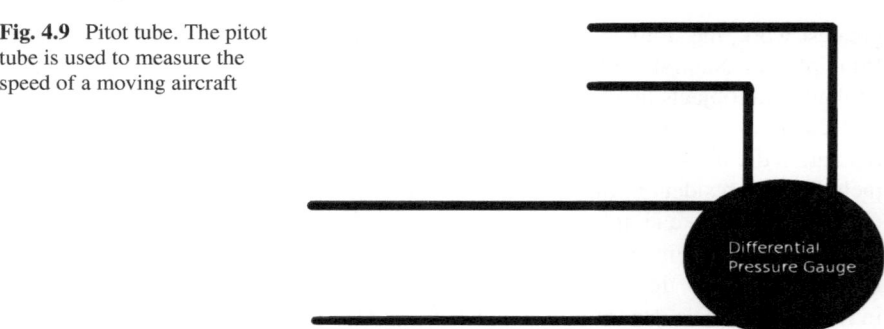

above it; hence the velocity was thought to increase with depth. Pitot report his stunning (and correct) finding that in reality the flow velocity decreased as the depth increased, thus introducing his new invention with flair."[7]

David Taylor devised a 13-hole Pitot tube to give him superior readings. Today the Pitot tube is used to measure air speed in wind tunnels and on aircraft. Whatever the fluid, the tube works by measuring the pressure inside the tube and comparing it to the static pressure outside the tube. When an object is sitting motionless in a still fluid, the pressure inside and outside the tube will be equal because the speed is zero. When the object is moving, there will be a difference in the pressure inside and outside the tube due to the motion of the craft. This difference can be used to calculate the speed of the craft. The same principle works to measure speed whether the medium is air, gas, or water.

Swiss mathematician Daniel Bernoulli (1700–1782) was a member of a well-respected family of scientists and mathematicians. It was with his book, "Hydrodynamica," published in 1783 that the term "hydrodynamics" made its literary debut. His fluid flow equation led to advances resulting in the modern paradigm of shipbuilding based on model design, a process pioneered by marine architects such as William H. Froude and David W. Taylor.

The entire design approach of Streamline Modern was inspired by Bernoulli's work. Popular in the 1930s, the aesthetic developed as a reaction to Art Deco and emphasized clean, aerodynamic lines as seen in airplanes and high performance vehicles. A prime example of this style is the 1934 Chrysler Airflow, which introduced fluid dynamically based engineering to the American auto market.

Today Bernoulli is famous for a principle that states that pressure decreases as velocity increases in a flowing field. (This is because the air moving over the top of a curved surface, airfoil, must move more quickly than the air moving beneath the same structure if it is to meet the streamline it moves with.) It's interesting to note that Bernoulli's famous equation is derived from Bernoulli's work, but Bernoulli did not actually formulate the equation in his lifetime. That was left to his contemporary, Leonhard Euler.

Leonhard Euler (1707–1782) is considered by many to be the preeminent mathematician of the eighteenth century. Euler lived in the same town as the Bernoulli's and went to study at the St. Petersburg Academy in 1725 with Daniel Bernoulli when he moved to Russia to teach and study. Euler worked closely with Bernoulli while in St. Petersburg, envisioning pressure as a point that could vary from point to point throughout a fluid. He created a differential equation for a fluid accelerated by gradients in pressure. He "conceived of pressure as appoint property that can vary from point to point throughout a fluid, and obtained a differential equation relating pressure and velocity."[8] The integration of that differential equation led him to create an equation based on Bernoulli's work. It is known today as the Bernoulli Equation.

Euler also developed equations based on his own work, which included discoveries in infinitesimal calculus that introduced the necessary precision for the formal study of fluid dynamics. *Euler's equations* for inviscid flows, those lacking friction, first appeared in his article "Principes generaux du mouvement des fluids" published in 1757. Euler's focus was on the fluid flow as a whole. The resulting analysis is of the fluid velocity at a fixed point. "Euler's contributions to theoretical aerodynamics were monumental; whereas Bernoulli and d'Alembert made contributions toward physical understanding and the formulation of principles, Euler is responsible for the proper mathematical formulation of these principles, thus opening the door for future quantitative analyses of aerodynamic problems—analyses that continues on to the present day."[9] Euler's equations are in use today and apply to compressible or incompressible fluids, depending upon the assumptions used for key variables.

"The successful derivation of these equations depended on two vital concepts that Euler borrowed in total or in part from previous researchers, as follows:

1. A fluid can be modeled as a continuous collection of infinitesimally small fluid elements moving with the flow, where each fluid element can change its shape and size continuously as it moves with the flow, but at the same time all fluid elements taken as a whole constitute an overall picture of the flow as a continuum [daVinci]…"[10]
2. Newton's second law can be applied in the form of a differential equation.[11]

"Utilizing the two concepts listed above, namely that of an infinitesimally small fluid element moving along a streamline, and the application of both the principle of mass conservation and Newton's second law to the fluid element in the form of differential calculus… Euler derived the partial differential equations of fluid motion that today carry his name and that serve as the foundation for a large number of modern aerodynamic analyses. The equations derived by Euler in 1753 *revolutionized* the analyses of fluid dynamic problems"[12] for inviscid fluid flows.

John Smeaton (1724–1792) was a British engineer who published a paper in 1759, "An Experimental Enquiry Concerning the Natural Powers of Water and Wind to Turn Mills and Other Machines Depending on Circular Motion." The theories in the paper were used for windmills, but it was its application to aerodynamics that held great significance for his work. His coefficient was especially of interest to those the Wright brothers and other early aviators wishing to apply science to the design of their flying craft. Unfortunately the Smeaton coefficient, which derived from his work and was not calculated by Smeaton himself, was incorrectly believed to be 0.005. Eventually the error was corrected, but along the way its use introduced error into a number of essential calculations.

Joseph-Louis Lagrange (1736–1813) was one of the creators of the calculus of variations. He also applied differential calculus to the theory of probabilities and proved that every natural number is the sum of four squares. One of his greatest contributions was to take the principles of Newtonian mechanics and, through the use of variational calculus, use them for analysis. This form of mechanics is now called Langrangian mechanics.

Lagrange also did work in the field of fluid mechanics. While Euler's work focused on the flow field, Lagrange looked at what is known as a *fluid parcel*—or finite area and volume in the flow. This focus results in an analysis of the trajectory of specific fluid parcels. The combined use of these approaches provides a thorough understanding of the characteristic of a fluid flow.

Fulton Through Taylor: Principles in Action

Robert Fulton (1765–1815) was an American who put hydrodynamic principles to practical use by being the first to successfully harness the power of steam, demonstrated by Hero, for profitable commercial purposes. His steamboats became a vital means of transportation for goods and people at the start of the nineteenth century when his North River Steamboat carried passengers between New York City and Albany, NY, and carried passengers upriver at the breathtaking rate of 5 mph. Fulton was not only involved in the use of steam for commerce.

In 1786, Fulton left the USA to study painting in England. While there, he worked on a project that involved the design of a canal system to replace the locks that were currently in use. He published a summary of his ideas on improvements to canal navigation in his "Treatise on Improvement of Canal Navigation," and headed to Paris for further research. Once in Paris, Fulton became fascinated by what was known as the "plunging boat." This was in actuality the submarine and was based on the ideas of an American inventor. Fulton's work led him to design not only the world's first steam warship, but the first practical submarine.

Claude Louis Marie Henri Navier (1895–1886) was the first to derive the governing equations for *viscous fluid flows*, those with internal friction. "However, … Navier had no concept of shear stress in a flow (i.e., the frictional shear stresses acting on the surface of a fluid element). Rather, he was attempting to take Euler's equations of motion and modify them to take into account the forces that act between molecules in the fluid."[13]

Sir George Gabriel Stokes (1819–1903), working independently of Navier and unaware of his work, also derived the governing equations of a *viscous fluid*. His dynamic viscosity coefficient appears in the Navier–Stokes equations. The Navier–Stokes equations apply to incompressible flows and must be modified with an energy equation when applied to compressible flows. Other than that adjustment, their equations remain unchanged to this day.

Theories were fine but British Civil Engineer Isambard Kingdom Brunel (1806–1859) was the colorful figure with enormous aspirations responsible for putting many of them to the test. All Brunel needed was to know it couldn't be done and he was off to face the challenge. It wasn't enough for him to construct a ship; he would construct the largest ship ever.

In 1833 he was appointed as railway engineer for the Great Western Railway. He oversaw construction of the line linking London to Bristol. This included the viaducts at Hanwell and Chippenham, the bridge over the Thames at Maidenhead, the 3,200-yard Box Tunnel outside Bath, and Bristol Temple Meads Station. An innovative businessman, it was he who moved to standardize the gage at 7-feet for purposes of high-speed and fuel economy. Brunel also devised the combination of a tubular, suspension and truss bridge to cross the Wye at Chepstow while working on the line from Swindon to Gloucester and South Wales. This design was improved further for the bridge over the Tamar at Saltash near Plymouth.[14]

In 1835 Brunel set out to build a transatlantic steamship service. The first ship would be the *Great Western* (Fig. 4.10). This ship was made of wood and used paddles propelled by the first steam engine in transatlantic service. In 1843 he launched the *Great Britain* (Fig. 4.11) with the first steam-powered passenger ship with an iron hull and screw propeller. His piece de resistance was the *Great Eastern* (Fig. 4.12). Designed with the cooperation of John Scott Russell and launched in 1858, it had a double iron hull and used both paddles and a screw propeller. Each of

Fig. 4.10 Great Western. Isambard Kingdom Brunel's *Great Western*. *Source*: Retrieved from http://commons.wikimedia.org/wiki/File:Great_Western_maiden_voyage.jpg

Fig. 4.11 Great Britain. Isambard Kingdom Brunel's *Great Britain*. *Source*: Retrieved from http://commons.wikimedia.org/wiki/File:SS_Great_Britain_with_four_masts_1853.jpg

these ships was the largest ship of its kind in the world at launch.[15] The *Great Eastern* held that designation for four decades but was not commercially successful. Most certainly a giant step up from Hero's original steam-powered orb or Fulton's paddle wheelers, *The Great Eastern* was used to lay the first Trans-Atlantic cable.

Fig. 4.12 Great Eastern. Isambard Kingdom Brunel's *Great Eastern. Source*: Retrieved from http://commons.wikimedia.org/wiki/File:Great_Eastern_1866-crop.jpg

John Scott Russell (1808–1882) was a British civil engineer. A founder of the Royal Institution of Naval Architects, he is best known for his design work in 1856 on the *Great Eastern*, the largest ship at the time, and in 1860 on the HMS *Warrior*, the world's first completely ironclad battleship.

His observations of waves in a shallow channel formed the basis for the theory of *soliton* in which the waves form a humplike formation (soliton) and work as a separate entity. The waves Scott Russell described were stable, traveling over long distances without dissipating. Their speed depended upon the size of the wave and when one wave met another they did not merge. His observations seemed to be exceptions to the work of Newton and Bernoulli so it took some time before soliton was understood.

His observations of water in a canal also led to the formulation of this wave-line theory. This theory would be cast aside in favor of Rankine's streamline theory, but for awhile it was the accepted explanation for the manner in which a ship moved through the water. In his presentation to the Institution of Naval Architects in 1860, Scott Russell tried to explain the motion of water and was described in this way: "Even as a matter of physical observation, of logical discussion, and of practical experiment merely, the author [Scott Russell] had found the motions of waves of water, produced by the disturbance of a ship, more difficult to understand thoroughly and clearly than any other subject of mechanical knowledge. And if hard to understand, it was much harder to explain. What becomes of the particles of water moved out of the way of a ship—where they go—how they get there—if they ever return to their old places—what force takes them away—what brings them back—if they don t come back, whence come those that replace them—how *they* come there, and how their place is in turn re-occupied?—all this requires minute observation of the phenomena before it can be understood."[16]

Scott Russell's explanation may sound ludicrous to us now, equipped as we are with the vision of hindsight backed up by two centuries of scientific exploration. But Scott Russell was unaware of streamlines. He also had no way of knowing that water passed under a vessel as it moved forward. "He told his hearers that, in steaming from England to America, a ship must excavate a canal three thousand miles long, and as large as the vessel's greatest section. He took it for granted that the power expended at the screw or paddles represented this work of excavation. To find the form of least resistance it was necessary to discover a shape that would move the water out of the way, just sufficiently to let the largest section pass and not a jot furthers, and that the ship finding the particles in her way at rest should leave them at rest in the new place to which they were moved."[17]

William Froude (1810–1879) was the first to champion and prove the validity of scale model testing in the design of ocean going vessels. He is also known for his pioneering work in the rolling of ships, helping Brunel to stabilize the *Great Eastern* and explaining for the first time the behavior of a ship experiencing waves and what could be done to minimize their impact. His *Froude Number*, still in use today, is a dimensionless number that measures resistance. It is calculated as the ratio of a body's inertia to gravitational forces. The greater the Froude Number, the greater is the resistance.[18]

One of Froude's greatest contributions was his Law of Similitude. According to this, the results obtained on a scale model would be applicable to a full-sized vessel or vehicle. This law is at the basis of all model testing. It is an integral part of the design process for vehicles and vessels of all types and led to the use of model basins worldwide for the testing of designs for the first metal ships.

William John Macquorn Rankine (1820–1872) was the Scottish civil engineer who developed the *streamline theory* that replaced John Scott Russell's wave-line system. The work in Rankine's paper, "On the mathematical Theory of Stream-lines, especially those with four Foci and upwards," published in 1871, was described in an abstract: "A stream-line is the line that is traced by a particle in a current of fluid. In a steady current each individual streamline preserves its figure and position unchanged, and marks the track of a filament or continuous series of particles that follow each other. The motions in different parts of a steady current may be represented to the eye and to the mind by a group of stream-lines.

Stream-lines are important in connexion with naval architecture; for the curves which the particles of water describe relatively to a ship, in moving past her, are stream-lines; and if the figure of a ship is such that the particles of water glide smoothly over her skin, that figure is a *stream-line surface*, being a surface which contains an indefinite number of stream-lines....

The author states that the occasion of the investigation described in the present paper was the communication to him by Mr. William Froude of some results of experiments of his on the resistance of model boats, of lengths ranging from three to twelve feet. ... In each case two models were compared together of equal displacement and equal length; the water-line of one was a wave-line with fine sharp ends [the Raven], that of the other had blunt rounded ends [the Swan], each joined in the midship body by a slightly hollow neck—a form suggested, Mr. Froude states, by the appearance of water-birds when swimming. At low velocities, the resistance of the sharp-ended boat was the smaller; at a certain velocity, bearing a definite relation to the length of the model, the resistances became equal, and at higher velocities the round-ended model had a rapidly increasing advantage over the sharp-ended model.

Hence it appeared to the author to be desirable to investigate the mathematical properties of stream-lines resembling the water-lines of Mr. Froude's bird-like models..."[19]

Rankine's work with streamlines was also part of the basis Froude used for his belief that scale model results would be applicable to full-sized ships, "... Now Professor Rankine's admirable stream-line investigations have definitely established the conclusion that for symmetrically shaped bodies of 'fair' lines, not excluding by that description certain very blunt-ended ovals, when wholly submerged, the entire resistance depends on the conditions of imperfect fluidity, of which surface-friction is the only one so considerable that we need take account of if we deal with bodies of rational dimensions; and this, as I have pointed out, does follow the law of the squares... If, therefore, we were dealing with submerged bodies, we should have no reason to mistrust the *prima facie* deductions founded on experiments with models...

The principles on which Professor Rankine's stream-line investigations are founded establish generally, in relation to all wholly submerged symmetrical bodies moving in a fluid infinitely extended on all sides, that the stream0line displacements which the motion of the body imposes on the surrounding volumes of fluid are, for a given body, identical in configuration for all velocities (an identity which assigns to the always a velocity proportional to that of the body itself), and that the configuration is similar for all similar bodies."[20]

Rankine also helped Froude refine his work on the rolling of ships. He was a founder of the science of thermodynamics, bringing heat into mathematical formulations for the first time.

While not a brilliant naval commander (he graduated second to last in his class at the Naval Academy and commanded several ships that collided with both moving and stationary objects), Alfred Thayer Mahan (1840–1914) wrote the book that changed the way navies were viewed worldwide. "The Influence of Sea Power Upon History" made the compelling case that throughout history, the nation with the strongest navy had had the power to protect her interests and ensure her place at the top of the power structure. He also drove home the point that the navies of the world were no long dependent upon the forces of the wind. They could move under their own power, choosing to battle at the moment and place that best suited them. "The power to assume the offensive, or to refuse battle, rests no longer with the wind, but with the party which as the greater speed; which in a fleet will depend not only upon the speed of the individual ships, but also upon their tactical uniformity of action. Henceforth the ships which have the greatest speed will have the weather-gage [advantage]."[21] Mahan's work was taught at naval colleges in the USA and abroad, influencing the major thinkers of his time. His theories sparked the first international race for naval prominence as navies strove to equip themselves with the fastest, most modern boats to gain the advantage described by Mahan.

Osborne Reynolds (1842–1912) put forth the *Reynolds Number*. In use today, it is a measure of the viscosity of a fluid. Developed by Reynolds in 1883, the Reynolds

Fig. 4.13 Suction in a
narrow channel. *Source*:
David Taylor's theory of
suction between two vessels
in a narrow channel was used
as evidence in the Olympic/
Hawke case of 1911

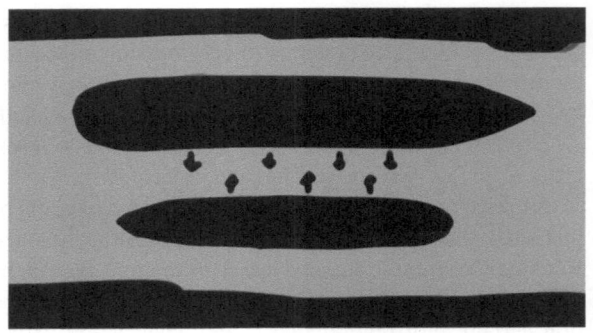

Number identifies the laminar flow (at low Reynolds numbers) with viscous forces
dominant, and turbulent flows (at high Reynolds numbers) where flow instabilities
are found. The Reynolds number is a basic parameter in the description of all fluid
flow situations, including the shapes of flow patterns, the ease of heat transfer, and
the onset of turbulence. The Reynolds number proved that the flow pattern over a
scale model would be the same for that of a full-scale version if the necessary flow
parameters were met in both cases.

David Watson Taylor (1865–1940) designed and operated the first experimental
model basin for scale model testing in the USA. Using Froude's groundbreaking work
as his springboard, he made modifications as necessary to accommodate the hot,
humid climate of Washington, DC. Rear Admiral Taylor's work at the Experimental
Model Basin was essential to the design of the US naval fleet of the twentieth century
and beyond. His work was instrumental in the adoption of the bulbous bow now used
by large vessels worldwide, and he performed tests that showed the streamlines moving past many of the models he tested.

His book, "The Speed and Power of Ships," first published in 1910 was a seminal
work. Still respected today, Taylor used what is known as the Taylor Standard Series
of model tests to arrive at the actual effect a change in set characteristics of a ship
would have on the speed and power of that ship. In this way he was able to use scale
models towed in a basin to estimate, accurately calculating the resistance of a ship
before it was built.

Taylor also studied the phenomenon of suction between two vessels moving
close by one another in a narrow channel. He was called as an expert witness in the
Olympic/Hawke trial after the cruise liner, RMS Olympic, collided with the British
battleship, HMS Hawke, off the Isle of Wight in 1911. It was determined that the
Olympic caused the accident because her larger displacement created greater suction and effectively "pulled" the Hawke toward her (Fig. 4.13).

A firm believer in the potential synergy of vessels and aircraft, Taylor was a
founding member of the National Advisory Committee for Aeronautics (NACA),
the precursor to the National Aeronautics and Space Administration (NASA), when
it was established by an act of Congress in 1915.

Each advance leading to the science of today required the work of these and other unknown, dedicated theorists who turned their single-minded devotion to the problems that intrigued them. The net result of their work is the body of scientific principles that constitute the field of hydrodynamics and inform the design of ships and other devices making their way through water for a variety of purposes.

Notes

1. (2012). "BBC History." 2012, from http://www.bbc.co.uk/history/.
2. Ibid.
3. Leonardo and B. Gates (2001). *Il Codice Leicester, Codice Hammer.* Napoli, A.E.S. Art Books.
4. (2012). "BBC History." 2012, from http://www.bbc.co.uk/history/.
5. Ibid.
6. Ibid.
7. Anderson, J. D. (1997). *A history of aerodynamics and its impact on flying machines.* Cambridge; New York, Cambridge University Press.
8. Johnson, R. W. (1998). *The handbook of fluid dynamics.* Boca Raton, FL, CRC Press.
9. Ibid.
10. Ibid.
11. Ibid.
12. Ibid.
13. Ibid.
14. (2012). "BBC History." 2012, from http://www.bbc.co.uk/history/.
15. Brunel, I. (1870). *The Life of Isambard Kingdom Brunel, Civil Engineer.* London, Longmans, Green, and Co.
16. Homans, I. S. (1860). *Hunt's merchants' magazine and commercial review*, Freeman Hunt: 804.
17. Ibid.
18. Gawn, A. (1955). An Evaluation of the Work of William Froude. *The Papers of William Froude, M.A., LL.D., F.R.S. 1810–1879.* R. N. Duckworth, Captain A. D. London, The Institution of Naval Architects.
19. London, T. R. S. o. (June 17, 1869 to June 16, 1870). *Proceedings of the Royal Society of London*, Taylor and Francis.
20. Froude, W. (1869). The State of Existing Knowledge on the Stability, Propulsion and Sea-going Qualities of Ships, and as to the Application Which it may be Desirable to Make to her Majesty's Government on this Subject. *The Papers of William Froude, M.A., LL.D., F.R.S. 1810–1879.* R. N. Duckworth, Captain A. D. London, The Instution of Naval Architects: 129–133.
21. Mahan, A. T. (1987). *The influence of sea power upon history, 1660–1783.* New York, Dover Publications.

Chapter 5
Aerodynamic Theorists

It is called Aerodynamics when the fluid is air.

Early theorists in the science of aerodynamics separate nicely into those who were heading for the sky and those who were intent on reaching outer space. The group that was intent on the moving through the sky can trace its origins back to a Greek philosopher and mathematician named Archytas. Their goal of attaining manned flight occurred with Wilbur and Orville Wright's success in 1903. The group intent on space can trace its origins back to the Chinese and their fire-rockets. They realized the possibility of space flight with the advent of Robert Goddard's rockets.

Unlike hydrodynamic advances that were largely a matter of observation, aerodynamic advances depended upon trial and error that often resulted in an untimely death for the innovator. Many of the early aerodynamic theorists will be familiar because of work with hydrodynamics discussed in the prior chapter.

Into the Air: Early Theorists

Archytas (428–347 BC) definitely had his mind on flight. He was a good friend of Plato and is thought to be the founder of mathematical mechanics. Five centuries after Archytas' death, Aulus Gellius wrote that Archytas had built a steam-powered model of a bird, "The Dove." The model was most likely suspended on a wire and propelled forward along the wire by the force of steam escaping from the rear of the model. The Dove is believed to have been the first self-propelled, artificial, flying device, although there is some disagreement as to how the bird was propelled and whether or not Archytas was actually the person who built it. Archytas is also the mind behind the harmonic mean in music, as well as the Archytas Curve—used to solve the problem of doubling a cube.

G. Hagler, *Modeling Ships and Space Craft: The Science and Art of Mastering the Oceans and Sky*, DOI 10.1007/978-1-4614-4596-8_5,
© Springer Science+Business Media, LLC 2013

Fig. 5.1 da Vinci's
parachute. da Vinci included
a sketch of a parachute in his
Codex Atlanticus

Leonardo da Vinci (1425–1519) next took up the pursuit of flight. He not only developed the continuity equation, as mentioned in the prior chapter, but also included a series of drawings of *ornithopters* in his *Codex on Flight*. These ornithopters were machines for flying that consisted of wings flapped by different types of mechanical devices powered by movement of the human arm, leg, or body. This mode of flight would prove to be impossible, since man cannot generate power to sustain his weight aloft. The fundamental differences between the anatomy of man and bird were not yet known, so those observing birds in flight assumed it was simply a matter of designing the right type of wing. Still, the ornithopter was an obvious attempt to mimic the flight of birds.

da Vinci also included a sketch of a strangely triangular parachute in his *Codex Atlanticus* (Fig. 5.1) but it was not until 1783 that John-Sebastian Lenormand of France would successfully parachute. While da Vinci may have considered a parachute a device to permit man to land like the birds, Lenormand viewed the parachute as a device that could be worn to escape fires on the upper floors of tall buildings.[1] Lenormand's first jump was from the top of a tree, using two parasols to slow his

Fig. 5.2 da Vinci's helicopter. da Vinci included a sketch of the forerunner to the modern helicopter in his *Codex on Flight*

rate of descent. His most famous jump was from the tower of the Montpellier Observatory in France. He jumped with a 14-foot parachute and landed unharmed.

In his *Codex on Flight*, da Vinci sketched a forerunner to a modern helicopter in which the blade was a vertical spiral (Fig. 5.2). He also sketched a glider (Fig. 5.3). Whether or not da Vinci used a glider himself is a subject of debate. Many point to his drawings from a "bird's eye view" as proof while others contend he could have gained that view from a high point. Either way, da Vinci's sketches were the first. It would not be until 1804 that Sir George Cayley of England flew the first successful glider model. In 1853 Cayley successfully flew a full-scale glider.

Another of da Vinci's enduring contributions is his theory that it wasn't necessary to move an object to measure the effects of movement on an object. In his *Codex Atlanticus* he wrote that the same measurements could be taken while having the fluid flow over a stationary object because the forces measured would be the same whether it is the object that moves through the air or the air that flows past the object.

Fig. 5.3 The da Vinci glider

This is the basis for the modern wind tunnel in which an object is mounted in place and the effects of flowing air past this stationary object are observed and measured.

Galileo Galilei (1564–1642) was an Italian physicist, astronomer, and mathematician. He played a major role in the Scientific Revolution and the development of the scientific method. Galileo believed in the heliocentric theory of the universe. Today we know this is correct but in 1633 this was not a widely held belief. Galileo stood trial for heresy. He was under house arrest for many years, during which time his works were banned from being reprinted. He continued to work during that time, even as he lost his sight.

Galileo's interest in astronomy led to his discovery of four of Jupiter's moons. Today they are known as the Galilean moons, in his honor. Galileo improved the magnification of the existing telescope of his time and was the first person to report finding craters on the Moon. Galileo also studied "kinematics," the study of uniformly accelerated objects. His work led to theories of parabolic trajectories, inertia, and the law of falling bodies, each one an important concept in the eventual design of machines that would make their way through the air.

Sir Isaac Newton (1642–1727) was an English physicist, mathematician, astronomer, natural philosopher, alchemist, and theologian. He is considered by many to be the greatest and most influential scientist who ever lived because of his work with gravity. Newton attributed his inspiration for the concept of gravity to watching an apple fall from a tree.

Newton added three laws of motion to the existing body of knowledge. These laws are the basis for several fundamental concepts of fluids. Because they have a significant place in aero as well as hydrodynamic principles, they bear repeating here. *Newton's First Law* is the law of inertia and states that a body in motion moves at a constant velocity covering equal distance in equal time in a straight line until acted upon by another force. Essentially this means that an object in motion will never stop unless acted upon by another force, and an

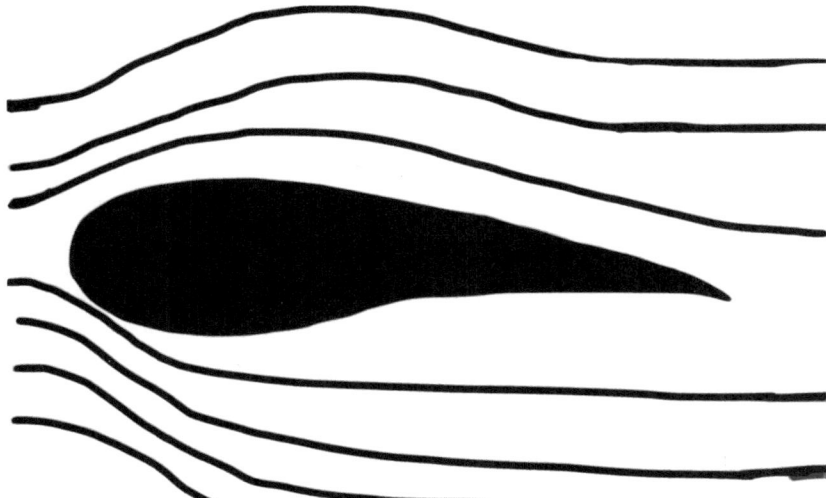

Fig. 5.4 Bernoulli's principle. The air moving more swiftly to cover the distance over the cambered surface of the wing results in an area of low pressure above the wing

object at rest will remain at rest unless acted upon by a force. *Newton's Second Law* states that the net force exerted on an object is equal to the product of that object's mass times its acceleration and takes place in the direction of the straight line in which the force acts. This means that you need more force to push a heavier object than you would to push a lighter object at the same acceleration. *Newton's Third Law* states that for every force there is an equal and opposite force. For a time this was thought to be the only component of lift.

The *Pitot tube* is a simple device first developed by Henri Pitot in the early eighteenth century. Originally designed to measure the speed of water, it is still used today to measure air speed in wind tunnels and on airplanes.

The importance of Daniel Bernoulli's (1700–1782) work to the field of aerodynamics cannot be overstated. It is the second form of his equation that explains the phenomenon of lift. Without an understanding of this principle, flight in a heavier-than-air craft would not be possible. Bernoulli's Law states that pressure in a flowing fluid field decreases as the velocity increases. This principle explains the phenomenon of *lift* on an airplane wing because when air flows past an airfoil, the air flowing over the top of the wing moves faster than the air flowing beneath the wing. The pressure on the top of the wing is then lower than the pressure beneath the wing, resulting in the lift that keeps an airplane aloft (Fig. 5.4).

The streamline theory is one offshoot of Bernoulli's work. Sir Horace Lamb (1849–1934) wrote in his book, "Hydrodynamics," "The preceding equations show that, in steady motion, and for points along any one stream-line, the pressure is, *coeteris paribus*, greatest where the velocity is least, and *vice versa*. This statement, though opposed to popular notions, becomes evident when we reflect that a particle passing from a place of higher to one of lower pressure must have its motion accelerated, and *vice versa*."[2]

Benjamin Robins (1707–1751) was an English military engineer who invented two testing devices. One was the *whirling arm* for measuring aerodynamic forces at low speeds. The other was a ballistic pendulum for studying aerodynamic characteristics of bodies at high speeds. The whirling arm was intended to provide a stead airstream for testing aerodynamic properties of a variety of objects. Through the use of the whirling arm and the ballistic pendulum, Robins verified Mariotte's finding that aerodynamic force varies with the square of the relative velocity between a body and the airstream. He also showed that two aerodynamic bodies with different characteristics but the same frontal area have different drag values. He was also the first to observe the Magnus effect and to note the increase in drag that occurs with speeds near the speed of sound.[3]

The work of Leonhard Euler (1707–1782) is also integral to both hydrodynamics and aerodynamics. His work with Bernoulli's findings led to the Bernoulli equation. This eponymous equation, created not by Bernoulli, but by Euler. Euler also introduced equations for fluid flow that could not be solved or applied in his time. This work provided a starting point for Italian mathematician Joseph Lagrange and French mathematician Pierre-Simon Laplace.[4]

John Smeaton (1724–1792) was a British engineer and the first Civil Engineer. In fact, he coined the term "Civil Engineer" to differentiate the type of work he did, designing and constructing structures like bridges, from other types of engineering work. In 1759 he published a paper, "An Experimental Enquiry Concerning the Natural Powers of Water and Wind to Turn Mills and Other Machines Depending on Circular Motion." The paper had applications to hydro and aero theorists but was especially of interest to those pursuing manned flight. His coefficient described the relationship between pressure and velocity. His conclusion was that they varied as the square of the velocity of the moving object. This coefficient was given the value of 0.005 and it was part of the reason that the Wright brother's expectations of lift were off on the 1900 and 1901 gliders, as we'll see in Chap. 7. The wind tunnel experiments conducted by the Wright's, as well as work by Langley and others, properly calculated the Smeaton coefficient at 0.0033.

Into the Air: Non-winged Flight

The Montgolfier brothers (1740–1810), Joseph-Michel and Jacques-Etienne, are credited with launching the first hot air balloon in 1782. One of 16 children of a successful French paper manufacturer, Joseph-Michel watched the embers rising from the fire in his fireplace one evening and deduced there must be some sort of gas with a property he called "levity." The gas he called Montgolfier Gas. He decided to

Fig. 5.5 Montgolfier
balloon. The Montgolfier
brothers flew the first
lighter-than-air object in 1782

use the gas to lift an object. His first attempts were with a simple box made of very thin wood and covered with taffeta. Once he saw that this was lifted to the ceiling when a fire was made beneath it, he worked with his brother Jacques-Etienne to build a balloon three times larger in scale. They successfully launched this balloon in 1782. It got away from them and flew just over a mile before landing. A crowd attacked it when it hit the ground.

This successful flight led the brothers to perfect their design and lay claim to their invention. Because they came from a papermaking family, they next built a balloon of sackcloth and lined it with three thin layers of paper. They flew this craft a few months later (Fig. 5.5).

They were then privileged to fly their balloon before King Louis XVI and Queen Marie Antoinette at Versailles. For this flight on September 19, 1783, live passengers would be aboard.

Since no one was certain what would happen to anyone flying at altitudes as high as 1,500 feet, they decided to fly three animals. A sheep was used because it was

Fig. 5.6 Balloon with animal test subjects. A sheep, duck, and rooster were sent aloft to determine what would happen to anyone flying at altitudes as high as 1,500 feet

thought their physiology was reasonably close to that of humans. A duck was aboard as a control. Because they were known flyers at those altitudes, any harm that came to them would be attributed to the influence of the craft itself. A rooster also went along because roosters were known to fly, but not at those altitudes. Because of this, the rooster would demonstrate any harm from the altitude itself. The animals went aloft and stayed aloft for about 8 min before landing safely (Fig. 5.6).

The Montgolfier brothers incorrectly believed they had discovered a new gas with a property called "levity." In reality the "new gas" was the buoyancy caused by the heated air inside the balloon. The air rose because of its decreased density in comparison to the air surrounding it. This was proven in 1785. For their ground-breaking work with lighter-than-air flight, Joseph and Jacques were honored by the French Academie des Sciences. They published books on aeronautics and Joseph-Michel invented both the calimeter and the hydraulic ram. Jacques-Etienne developed a process for the manufacture of vellum.

Pilatre de Rozier (1754–1784) and the Marquis d'Arlandes (1724–1809) were the first people to leave the ground and remain above it for an extended period of time. They did this in a hot air balloon designed by the Montgolfier brothers. Their flight in 1783 lasted 25 min, covered 5 miles across France, and proved that people could move through the air at the same altitude as the birds.

As a result of his successful ride, Pilatre de Rozier decided to make his own balloon. Unfortunately, this balloon contained both hot air and hydrogen. Rozier and a friend decided to fly from Boulogne to England. On June 15, 1785, at an altitude of about 3,000 feet, the hydrogen in the balloon exploded after being expanded by the hot air. The two men were killed.

Louis-Sebastien Lenormand (1757–1837) was not only the first to make a witnessed parachute decent, he is the one who created the word "parachute." (*Para* is Greek for against. *Chute* is French for fall.) This Frenchman first jumped from a treetop, using two parasols to slow his rate of fall. His next attempt, on December 26, 1783, was from the tower of the Montpellier Observatory. Quite a crowd, including Joseph Montgolfier, gathered to witness his jump with a 14' parachute with a rigid frame. The jump was successful, although it seems Lenormand's interest in the parachute was as a means of escaping fire in a tall building rather than something that would have use in the pursuit of flight.

Andre-Jacques Garnerin (1769–1823) was the first to jump from a high altitude with a nonrigid parachute. He made his first jump in October of 1797 by going aloft, attached to a balloon in place of a carriage (Fig. 5.7). His parachute was closed until the moment of release and the air inflated it as he fell. Garnerin later wrote, "I was on the point of cutting the cord that suspended me between heaven and earth… and measured with my eye the vast space that separated me from the rest of the human race… I felt myself precipitated with a velocity that was checked by the sudden unfolding of my parachute."[5] Garnerin made jumps from hot air balloons as high as 8,000 feet. He also designed vents in his parachutes to reduce oscillation as he fell.

In 1785, Jean-Pierre-Francois Blanchard (1753–1809), an avid French balloonist crossed the English Channel by balloon. American physician John Jeffries accompanied him on this first aerial crossing of the Channel. They lost altitude along the way and were forced to jettison everything but the mail they carried; the first airmail ever delivered. Blanchard went on to make ballooning demonstrations in several countries, including America. At the American ascent in 1793, President George Washington was an observer.

Into the Air: Winged Flight

In 1799, Sir George Cayley (1773–1857) designed the first airplane that incorporated a fixed wing, a separate propulsion mechanism, and a tail for stability. He engraved his concept on one side of a silver disk. On the other side he engraved a diagram of the forces that would be in play to produce lift on a fixed wing. "The arrow shows flow from right to left, and the heavy diagonal line represents a wing cross section at a rather large angle of attack to the flow. In the right triangle above the wing, we see that the hypotenuse represents the resultant aerodynamic force, and the horizontal and vertical sides represent the drag and lift, respectively"[6] (Fig. 5.8).

Cayley's design was revolutionary because he broke *lift* and *propulsion* into two parts, unlike the designers of *ornithopters* who expected the wings would provide

DESCENTE DE JACQUES GARNERIN
EN PARACHUTE (1797)

Fig. 5.7 Garnerin. Garnerin was the first to hitch a ride on a hot air balloon, then cut the cord and parachute safely to earth. *Source*: Retrieved from http://commons.wikimedia.org/wiki/File:Early_flight_02561u_(4).jpg

Fig. 5.8 Cayley's coin. Sir
George Cayley engraved the
four forces of flight and his
concept for a fixed-wing
flying machine on a coin

both lift and propulsion. His design introduced the airplane configuration we use today and sparked the race for manned flight in a vessel other than a balloon.

Cayley was also the first to introduce *camber* into airfoil design. Up until his time, wings were flat. As a result of this, they could not produce lift under all circumstances. Cayley recognized that a wing with a curve to it had distinct advantages over a flat wing. His cambered airfoils became the standard.

Octave Chanute (1832–1910) was born in France, moved to the USA and became a citizen, settling in Chicago. A well-respected engineer, he became interested in flight in 1875 and designed several biplane gliders. Because he was in his 60s at the time, he did not personally fly his gliders although he was there to observe each flight. "He corresponded with and was respected by virtually all of the principal workers in aeronautics at the time. He served as a catalyst, inspiring and encouraging others in their efforts toward powered, manned flight. His book was read by the Wright brothers…"[7] In 1894 his book, "Progress in Flying Machines," was published. "…it was the definitive publication to date on the history and current status of flying machines: 'Eighty years after the original publication, *Progress in Flying Machines* remains one of the most comprehensive and reliable histories of pre-Wright aeronautics available.'"[8]

Wilbur Wright wrote to Chanute in 1900 and their correspondence continued until Chanute's death in 1910. Chanute was an invaluable sounding board for Wilbur Wright. While the brothers did not accept any money for the pursuit of their goal, they did use Chanute's anemometer and allow some of his men to accompany them to Kitty Hawk. Chanute was also a welcome visitor at Kitty Hawk.

Samuel Pierpont Langley (1834–1906) had a distinguished reputation in astronomy by the time his interest turned to aviation. He had achieved fame for his work with sunspots, invented the radiometer to measure the distribution of heat in the solar spectrum, developed a method to determine the solar constant of radiation, and was about to be named the third secretary of the Smithsonian Institution (1891) where he would establish the Astrophysical Observatory.

Langley's work was funded by grants from the Smithsonian and the War Department. He began with a small series of twisted rubber band- propelled models to test his theories. Unlike Penaud, Langley was unable to achieve steady sustained flight with these hand-held craft. Using a whirling arm at the Smithsonian, Langley experimented with many combinations of aircraft components. He was convinced that a practical aircraft should be inherently stable. This allowed him to neglect the

Fig. 5.9 Langley 01. Langley's Aerodrome A sits atop the houseboat he used for launching. *Source*: NASA/courtesy of nasaimages.org

Fig. 5.10 Langley 02. An aerodrome just after launch. *Source*: Retrieved from http:// commons.wikimedia.org/ wiki/File:Samuel_Pierpont_ Langley_-_Potomac_ experiment_1903.jpeg

matter of flight control on his models. He also believed the same degree of stability could exist in full-sized, piloted models. With the concept Langley had in mind, the pilot would essentially have little to do but enjoy the ride.

Langley chose to fly his planes over the Potomac River to reduce the risk of catastrophic injury to his planes or pilots. To get his planes in the air, he used a catapult that sat atop a houseboat, an alternative that subjected his craft to considerable stress (Fig. 5.9 and Fig. 5.10). Ultimately his unpiloted planes were the first powered

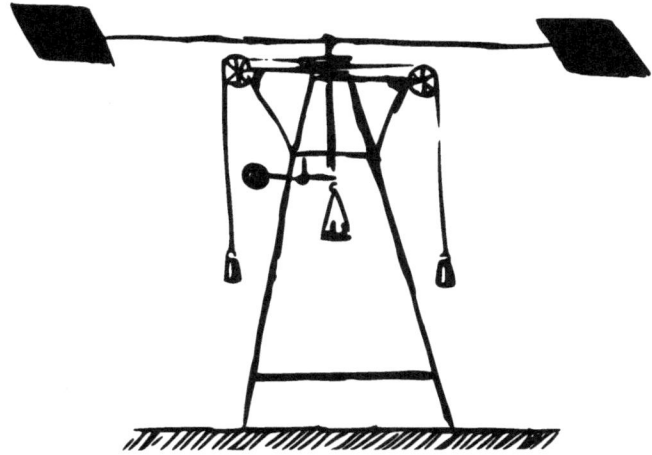

Fig. 5.11 Lilienthal Whirling arm. Lilienthal used his whirling arm to test airfoils before building his gliders

heavier-than-air machines of significant size to achieve sustained flight, as witnessed by Alexander Graham Bell. But his piloted flights were another story.

He unsuccessfully attempted piloted flights two weeks before the Wright brothers successful flight in 1903 and it was feared his lack of success would cast a shadow on his otherwise notable career. By the time Langley finished his work with unmanned and then manned gliders, he was the object of derision. His theories had not proven successful, despite the large amount of funding he had received and the brilliance he'd displayed in all other aspects of his career.

Meanwhile, in his native Germany, Otto Lilienthal (1848–1896) was achieving international fame as the first to build and fly a controlled glider. He based his designs on experiments he performed on a *whirling arm* device he constructed for that purpose (Fig. 5.11). Many of the airfoils he tests were cambered and he became convinced these curved surfaces were the most aerodynamically efficient. He made history in 1891, with a successful flight made with his glider attached to his shoulders. The glider was controlled by Lilienthal's movements but this gave him limited mobility for maneuvering and was a chief cause of his premature death in 1896 due to a crash when a wind gust upset the delicate balance required to maintain flight.

It was Horatio Frederick Phillips (1845–1926) who first demonstrated the lift described by Cayley. Phillips was an English enthusiast who patented eight wing-like sections of various widths and curvatures in 1884. He used a "wind box" to calculate the required velocity of the oncoming stream of air to generate lift for a given weight. His experiments proved that a cambered surface creates more lift than a flat surface.[10] The "wind box" used by Phillips was only the second wind tunnel ever used. His use of the controlled airstream of the wind tunnel allowed him to achieve reliable results and move the theory of aerodynamic forces forward by establishing the validity of the cambered surface as the most aerodynamically efficient.

"In 1891, Phillips devised and patented an improved wing section designed to create even more lift. He explained that low pressure is produced on the blade's upper surface, while high pressure is produced on the underside. Since high pressure always moves toward low pressure, the high pressure below pushes the blade upward to the low pressure and creates lift. In 1893, he created a 350-pounds (158.8-kilograms) model aircraft that ran around a 628-feet (181.4-meter) circular track attached to a central pole. The model rose about three feet (91 centimeters) off the ground when it reached a speed of 40 miles per hour (64 kilometers per hour). This model had fifty rows of superimposed small winglets arranged in a slat-like fashion on wheels. Each slat was twenty-two feet (6.7 meters) long and 1.5 inches (3.8 centimeters) wide and was mounted two inches from the next slat. A coal-fired engine turned a twin-bladed propeller 400 revolutions per minute."[11]

The next to advance the theory of aerodynamic forces was American inventor Sir Hiram Maxim. He performed his tests with both a whirling arm and a wind tunnel with the goal of creating a machine capable of rising in the air. His huge steam-powered biplane was to be the test subject. "It was the largest flying machine ever built up to that time. The four wheels of the machine rested upon straight rails that were 1,800 feet (548.6 meters) long. He used the rails to launch his giant steam-powered biplane and also to prevent it from escaping its test track and climbing into uncontrollable flight. The 'flights' began in early 1893. Although the machine could not really fly, it lifted up off the ground and shot forward more than 1,800 feet (548.6 meters). On July 31, 1894, in what was to be the last of its experiments, the machine broke loose of one of its rails while traveling at 42 miles per hour (67.6 kilometers per hour). However, the free 'flight' did not last long. A piece of the broken guardrail hit the propeller, and Maxim shut off the steam. Maxim demonstrated that a powerful engine could lift a heavy winged object from the ground."[12] This demonstration was a milestone in aerodynamics.

Percy Pilcher (1866–1899) was another early aviator with an interest in gliders so great this British innovator built a number of them. His first was the Bat, built in 1895. Wishing to know more about glider design, Pilcher met with Glider King Otto Lilienthal. Pilcher's future gliders included several of Lilienthal's configurations and techniques. One in particular was the manner in which Lilienthal "wore" his glider on his shoulders and flew suspended from the glider as though beneath a very large kite. As a result of that meeting, Pilcher built and flew The Hawk in 1896. This was Pilcher's most successful glider. With The Hawk, Pilcher was able to break the world distance record for unpowered, piloted flight.

Pilcher also corresponded with Octave Chanute about the design of a powered flying machine when he had difficulty designing a plane that would generate enough lift to carry the passenger and motor, yet be of reasonable dimensions. (This was because the greater lift he needed required too large a wing.) At Chanute's suggestion, Pilcher solved the problem of lift versus wingspan by designing a triplane (Fig. 5.12) with small, light wings stacked atop each other to generate the

Fig. 5.12 Pilcher. Pilcher's triplane had small, light wings stacked one above the other

lift sufficient to carry the load aloft. On the day he crashed, the triplane was complete but needed repair, so Pilcher instead chose to use The Hawk for a demonstration of flight for potential sponsors. The tail snapped during that flight and Pilcher crashed, dying 2 days later without ever having flown his triplane.

French experimenter Clement Ader (1841–1926) was active in gliding experimentation during the same time period as Percy Pilcher. His earlier investigations took place in 1870 when he constructed a balloon during the Franco-German War. By 1876 he'd quit his job to make more money to support his aviation avocation through the creation of electrical communications devices such as the microphone and a public address device.

By 1890 his focus was on heavier-than-air flight. It was at this time Ader constructed a steam-powered, bat-winged monoplane. Named it the Eole for Greek god of wind. Ader's bat glider included heavily cambered wings. On October 9, 1890, Ader "flew" his glider a few inches off the ground. It was the first steam-powered craft to rise from the ground but it did not have a piloting system and could not sustain flight.

In 1897, the French War Ministry commissioned Ader to build a new plane for testing. The Avion III was again driven by steam and again lacked a means for control once aloft. Tests of the plane proved unsuccessful; it did not take off and instead ended up in a field.[13]

Enter Orville and Wilbur Wright. In 1901, the Wright brothers (1867–1948) were dispirited by the poor performance of their glider and well aware of the work done by Cayley, Chanute, Langley, Lilienthal, and Pilcher when Wilbur made the statement that no one would fly in his lifetime. Despite that, Wilbur, along with his brother Orville, brought a singular determination to the problem of manned flight in a heavier-than-air craft. The Wright brothers were the first to test airfoils in a wind tunnel and use the results to inform the design of future craft when their experience led them to doubt the science of the day. They tested airfoil designs for gliders to arrive at an airfoil design that would be successful. Soon enough, they turned their attention to powered flight. Their persistence in flying a heavier-than-air structure ignited the imagination of the world when their powered, manned craft took off and flew at Kill Devil Hill near Kitty Hawk in North Carolina on December 17, 1903.

By the 1920s, flight was proven and the focus shifted to distance and speed. American aviator Charles Lindbergh (1902–1974) was the first person to fly solo over the Atlantic Ocean. He made this flight in 1927. Lindbergh was quite eloquent about his time in the sky, remarking about his first parachute jump in *The Spirit of St. Louis* in 1953, "It was a love of the air and sky and flying, the lure of adventure, the appreciation of beauty. It lay beyond the descriptive words of men—where immortality is touched through danger, where life meets death on equal plane; where man is more than man, and existence both supreme and valueless at the same time."

Amelia Earhart (1897–1937), an American, was the first woman to cross the Atlantic. She made the flight in 1923. In 1932, she made a solo flight across the Atlantic. She disappeared while on an around-the-world flight in 1937. She was no less eloquent on the reasons for flight than was Charles Lindbergh, "After midnight the moon set and I was alone with the stars. I have often said that the lure of flying is the lure of beauty, and I need no other flight to convince me that the reason flyers fly, whether they know it or not, is the esthetic appeal of flying."[14]

Ludwig Prandtl (1875–1953) was a German professor of mechanics at the Technical Institute of Hanover. Deservedly referred to as the father of aerodynamics, he was named director of what is now the Max Planck Institute for Fluid Mechanics in 1925. Prandtl identified the *boundary layer* in 1904 and carried out experiments on airflow over airplane wings. He made lasting innovations in the design of wind tunnels and other equipment related to aerodynamics. His work with streamlining airships and his support of monoplanes led to advanced heavier-than-air aviation exploration.

It's difficult to appreciate the importance of the boundary layer theory. This one theory elegantly incorporates streamlining and drag into an understanding of skin friction drag. His paper, "Ueber Flussigkeitsbewegung Bei Sehr Kleiner Reibung" (Fluid Flow in Very Little Friction) described his boundary layer theory in which "there was an extremely thin of fluid around a wing or airfoil that stuck to it because of friction. The friction caused this thin layer of fluid, called the *boundary layer*, to move, or flow, around the wing very slowly as if it were being dragged or pulled over the surface. The farther away from the wing's surface the layer of air was, the less it was affected by friction and the faster it moved until it reached the outer edges of the boundary layer, where the airflow was normal and the fluid moved at normal speed.

Professor Ludwig Foppl commented on Prandtl's boundary layer theory in his memoir: 'In view of the importance of this work, I would like to point out to its essentials. By that time, there had been no theoretical explanation for the drag experienced by a body in a flowing liquid or in the air. The same applies to the lift on an airplane. Classical mechanics was either based on frictionless flow, or, when friction was taken into account, mathematical difficulties were so enormous that hitherto, no practicable solution had been found. Prandtl's idea that led out of this bottleneck was the assumption that a frictionless flow was everywhere with the exception of the region along solid boundaries. Prandtl showed that friction, however small, had to be taken into account in a thin layer along solid walls. Since that time, this layer has been known as Prandtl's boundary layer. With these simplifying assumptions, the mathematical difficulties just mentioned, that show up in classical fluid mechanics of a flow with friction, could be overcome in a number of practical cases. Prandtl could prove theoretically and experimentally that the boundary layer can separate from the surface of a body immersed in a flowing fluid at suitable points, to roll up and leave the body as an isolated vortex.'

Prandtl also observed that flow separation was another possible result of friction. When a certain type of flow occurred, the boundary layer separated from the surface of the wing. This resulted in a region of slow-moving air behind the wing. This slow-moving air had lower pressure than the air flowing over the front of the wing. This change in pressure distribution around the wing resulted in a pressure drag toward the rear of the aircraft that much exceeded friction drag."[15] Prandtl's theories are still in use today.

Into Space

Those interested in reaching the stars put their interest in rockets. The history of rocketry can trace its origins to the first century AD when the Chinese began experimenting with gunpowder and gunpowder-filled tubes. The Chinese eventually discovered they could tie the tubes to arrows and shoot them at their enemies. With this innovation, the rocket was born, although it wasn't until Robert Goddard's liquid propellant rockets in 1926 that their full promise became evident.

It was Roger Bacon (1214–1292), a Franciscan friar living in Britain, who improved on the work of the Chinese fire-rocket engineers by improving the recipe for gunpowder. He included this recipe in his work, "The Epistola Fratris R. Baconis." His recipe is believed to have been a vast improvement over earlier formulations. Bacon was also a politician. By 1618 he was appointed lord chancellor, the most powerful position in England. In 1621 he was created viscount St. Albans. Unfortunately, he was charged by parliament with accepting bribes shortly after that and was fined, imprisoned, and banished from the court until the king pardoned him.

Joanes de Fontana of Italy designed and built the first torpedo powered by rocket in 1420. It ran at the surface and was launched and aimed at enemy ships. It is said they lit the enemy vessels on fire.

An Italian mathematician, Nicolo Tartaglia (1500–1557) wrote on the application of mathematics to artillery fire in his book, "Nova Scientia." Published in 1537, the book was an attempt to establish the laws of falling bodies and detailed new instruments and methods for his new branch of science: *ballistics*. Because of his groundbreaking work he is referred to today as the *father of ballistics*.

The first multistage rocket design is credited to Kazimierz Siemienowicz (1600–1651). This commander in the Polish Royal Artillery was an expert in artillery and rocketry. His manuscript on rocketry, "Artis Magnae Artilleriae pars prima," was partially published before his death. (There is speculation he was murdered by guild members who wanted to keep what he wrote about secret.) His book included the design for a multistage rocket and it was this design that became the basis for rocket technology for rockets launched into space. Siemienowicz's book included instructions for creating rockets, fireballs, and other pyrotechnic events.

Colonel William Congreve (1772–1828) was the first to launch rockets from ships. Some of his designs had ranges of 6,000 yards. He created case-shot rockets to spray the enemy with carbine balls. To burn ships and buildings, he created incendiary rockets. He also adapted some of the projectiles for land combat. While the rockets could outrange guns of the time, they were exceedingly inaccurate. They also were prone to exploding prematurely. Still, the loud explosion, red glare, and occasional hit made these rockets a source of concern for the enemy and reportedly had a demoralizing effect on them.

Jules Verne (1828–1872) was a French science fiction writer and visionary. In one of his books, "De la Terre de la Lune," (From Earth to the Moon), a giant cannon fired a manned projectile at the Moon. The projectile, launched from Florida, was named *Columbia*. Not a rocket, but an actual space ship, it carried a crew of three who experienced weightlessness while in space. Obviously, there was no way Verne could have known about weightlessness, or about the efficiencies to be had by launching a space vessel headed to the Moon from location in Florida, yet both were included in his novel.

Verne was a proponent of scientific investigation, as seen in this quote from "Journey to the Center of the Earth," "Science, my lad, is made up of mistakes, but they are mistakes which it is useful to make, because they lead little by little to the truth." Verne was a proponent of the benefits of collaborative research and the sharing of results. "Anything one man can imagine, other men can make real," he wrote in "Around the World in 80 Days."

Verne wrote in "De la Terre de la Lune," "In spite of the opinions of certain narrow-minded people, who would shut up the human race upon this globe, as within some magic circle which it must never outstep, we shall one day travel to the moon, the planets, and the stars, with the same facility, rapidity, and certainty as we now make the voyage from Liverpool to New York!" Because of the viewpoint he expressed, Verne's books brought travel to the Moon and the thrill of discovery alive in the popular imagination.

Alphonse Penaud (1850–1880) was a Frenchman who invented the first heavier than air, powered model to fly through the air. It was powered by a piece of rubber. He first flew his Planophore in Paris in 1871. He employed a *"cruciform tail"* which became known as the Penaud tail. He also had an ornithopter powered by rubber. These planes were very popular toys for kids and the Wright brothers played with one as boys (Fig. 5.13). Penaud's glider was the first aircraft to exhibit longitudinal

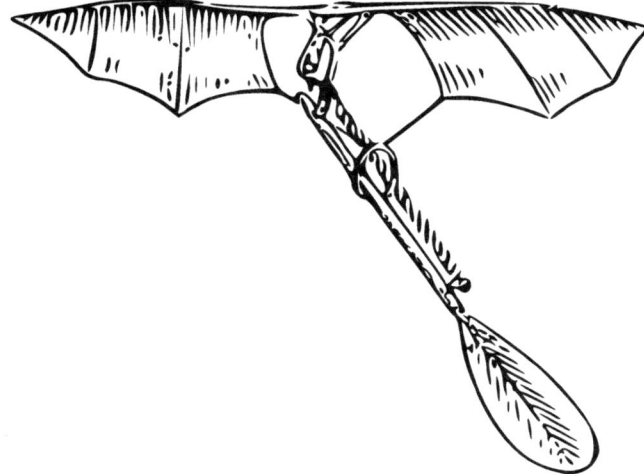

Fig. 5.13 The Penaud Mechanical Flyer

stability. He went on to design a full-sized glider that he never built or flew. His location of the wing and pitch of the tail on his glider were revolutionary and in keeping with the configuration of modern airplanes today.

"Pénaud next developed a two-passenger, full-size amphibian monoplane with his mechanic Paul Gauchot. He applied for the patent in 1876, but the model was never built. This two-seater had several features that would appear in future aircraft: double elevators, and a rudder connected to a fixed vertical fin, counter-rotating propellers, a glass-domed cockpit, retractable landing gear with shock absorbers, and piloting instruments. The estimated weight was to have been 2,635 pounds (1,195 kilograms), and the speed 60 miles per hour (96.5 kilometers per hour). His designs never came to fruition as he committed suicide in 1880 at the age of thirty."[16]

Konstantin Tsiolkovsky (1857–1935) is considered the *father of astronautics, cosmonautics, and human spaceflight*. He was a proponent of rocket engines powered by liquid propellants, orbital space stations, energy from the sun, and colonization of the Solar System. He published his most famous work, "Research into Interplanetary Space by Means of Rocket Power," in 1903, the same year of the Wright brothers' first flight. His rocket equation was based on Newton's Third Law.

Tsiolkovsky's interests included all aspects of the air around him. "The blue distance, the mysterious Heavens, the example of birds and insects flying everywhere—are always beckoning Humanity to rise into the air," from "The Successes of Air Balloon in the XIX Century," 1901.

"In 1926, Tsiolkovsky published, a bold 16-step program whereby human civilization could outlive its dying sun and settle the universe. The scheme called for rocket-powered airplanes, the use of plants for life support, and solar radiation to grow food and supply energy. He predicted the need for spacefarers to use pressurized suits when leaving the spacecraft, and envisioned the construction of large orbital settlements. According to Tsiolkovsky, humans would colonize the asteroid belt, the solar system, and ultimately the galaxy.

That work was followed three years later by 'The Space Rocket Trains,' which advanced Tsiolkovsky's earlier thoughts about multistage rockets. His calculations proved that building a rocket with separate stages, each of which would be jettisoned as it finished consuming its propellants, would allow a payload to be accelerated indefinitely.

Tsiolkovsky's publications are full of ideas that would later become common practice in aerospace engineering. He proposed using graphite rudders to steer a rocket in flight, cryogenic propellants to cool combustion chambers and nozzles, and pumps to drive propellant from storage tanks into the combustion chamber. He considered human factors as well—at the dawn of the Space Age, the first cosmonauts were amazed by the accuracy of Tsiolkovsky's descriptions of life in weightlessness."[17]

Robert Hutchins Goddard (1882–1945) was an American scientist and professor who flew the world's first liquid propellant rocket in 1926. It only climbed 41 feet but it was the forerunner to the Saturn V Moon rocket. Goddard is often referred to as the *father of modern rocketry* and, among other things, developed a gyroscopic system to control his rockets while in flight. He was passionate about his work and prototypes, introducing a parachute recovery system for his rockets. Goddard also filed for a number of patents, including one for "pluarity," Goddard's term for multistage rockets.

The work of these "rocket men," each laboring in their own country without full knowledge of the work of the others, moved the state of science to the point where rocket-powered vessels were proven as a viable way to move beyond our atmosphere and into space.

Notes

1. (2011). Louis-Sebastien Lenormand. *Encyclopedia Britannica*, Encyclopedia Britannica.
2. Lamb, S. H. (1895). *Hydrodynamics*, University Press.
3. Anderson, J. D. (1997). *A history of aerodynamics and its impact on flying machines.* Cambridge; New York, Cambridge University Press.
4. "Centenniel of Flight."
5. Garnerin, A. (1802). M. Garnerin's Account of His Ascent from St. George's Parade, North Audley Street, and Descent with a Parachute, Sept 21, 1802. *The European Magazine and London Review.* **42**.
6. Anderson, J. D. (1997). *A history of aerodynamics and its impact on flying machines.* Cambridge ; New York, Cambridge University Press.
7. Ibid.
8. Ibid.
9. Ibid.
10. "Centenniel of Flight."
11. Ibid.

12. Ibid.
13. Ibid.
14. Earhart, F. o. A. "Amelia Earhart." 2012, from http://www.ameliaearhart.com.
15. "Centenniel of Flight."
16. Ibid.
17. Sergeeva, G. a. T., Elena (2012). "Russian Space Web." 2012, from http://www.russianspace-web.com.

Part III
Scale Model Testing Begins

True innovators are hard to find. They are the ones who conceptualize a new or different way to accomplish a task and set out to make it happen. They are often ridiculed by those who find their ideas threatening for reasons ranging from the existence of a personal stake they may have in the matter to an inability to move from the known. Whatever the cause, a true innovator needs more than the courage of his convictions. He needs the ability to tune out the naysayers and get on with his work.

True innovators are those with the ability to take what is available, modify it when it can be used in a new way, scrap it when it cannot, and invent new ways when none is otherwise available. This ability is what led William Froude to insist that scale models would have value for design of full size vessels. It is what guided David Taylor in his quest for the ultimate testing facility. This ability and willingness to try something new led early aviation pioneers to devise new machines and, when necessary, to launch themselves into the way of bodily harm. The Wright brothers carried on this legacy, devising a way to test what they thought were unreliable results and taking part in experiments that involved their personal safety. The Rocketmen involved in the infancy of rocketry were equally adept at devising methods for testing their theories about the behavior of rockets under conditions they had never personally experienced.

The history of endeavor can be a dry recitation of facts or a journey with the innovators. The next chapters take you along as the use of scale models revolutionized the manner in which large vessels are designed.

Chapter 6
William Froude

> *I contend that unless the reliability of small-scale experiments*
> *is emphatically disproved, it is useless to spend vast sums of*
> *money upon full-sized trials, which, after all, may be*
> *misdirected, unless the ground is thoroughly cleared beforehand*
> *by an exhaustive investigation on small scale.*
>
> William Froude, 1868

For centuries ships were constructed of wood, powered by sail, and built to specifications that had worked in the past or "should" work going forward. Informal tests of ship designs had been made with simple models from time to time. Since none had provided results that were reliable indicators of the performance of the full-scale ships when completed, it was widely accepted that model tests would never prove useful. It wasn't until the introduction of three grand vessels by I.K. Brunel at the start of the nineteenth century that the notion of model testing came up for renewed debate.

Brunel's *Great Western* (1837), *Great Britain* (1843), and *Great Eastern* (1858) were each the largest ship at the time of her launch date. Their new iron hulls broke the existing constraints on ship length imposed by wooden hulls, even as their paddlewheels and screw propellers replaced sails and introduced new constraints on hull design. When the *Great Western* not only proved Brunel's theory that it took proportionally less fuel to move a larger ship than a smaller one, but also established the fact that accurate power estimates were possible, naval engineers were in a quandary. How could they take full advantage of these expensive and time-intensive new technologies when they had no history to guide them?[1]

The time was right for a reexamination of scale model testing when English civil engineer and visionary William Froude stepped forward in 1868 to champion the use of scale model testing in the design of ships before construction began. No one at the time imagined the far-reaching and historic impact Froude's proposal would have in the next centuries.

G. Hagler, *Modeling Ships and Space Craft: The Science and Art of Mastering*
the Oceans and Sky, DOI 10.1007/978-1-4614-4596-8_6,
© Springer Science+Business Media, LLC 2013

William Froude

William Froude had an avid interest in the science of ships. The Oxford-educated mathematician's interest was life long and took the form of observation and direct inquiry. He had worked for I.K. Brunel on the Great Western Railway, suggested bilge keels to successfully reduce the rolling of Brunel's *Great Eastern*, made some rudimentary self-propelled scale model trials on Dartmouth Creek, and submitted several well-respected papers to the newly formed Institution of Naval Architects before he approached the British Admiralty in 1868. With the backing of Chief Constructor of the Navy Edward J. Reed, Froude proposed the formal use of scale models in the testing of designs for full-sized ships. The models would be used to arrive at the optimal dimensions for adequate speed given the realities of the resistance and rolling of ships made of iron, powered by steam, and driven by screw propellers when moving through open water. To build a facility and perform these tests, Froude requested funding for an experiment tank he would build and run near his home in Torquay, England.[2]

The fact that members of the Institution of Naval Architects were familiar with Froude and the excellent quality of his work from his seminal work on the rolling of ships did not stop their vehement reaction against the use of models for such important purposes. What model tests had been done before had not been done on a formal basis. Institution members were certain the test results would be inaccurate for two principal reasons: The first was their belief that waves made by models were proportionally larger than waves made by a full-sized ship. The second was their certainty that the viscosity of water was a more important factor when dealing with a model than with a full-sized ship.

Those arguing against models preferred to test completed, full-sized ships and use the results of those tests for the design and construction of future ships for the Royal Navy. In fact, C.W. Merrifield, Esq., F.R.S, Principal of the Royal School of Naval Architecture and Marine Engineering, had submitted a plan of his own "to conduct experiments upon Her Majesty's ships in the fiords of Norway or on the inland waters of the West Coast of Scotland." When his plan was ultimately rejected in favor of Froude's model tests, he said, "although I still adhere to my preference for experiments on full scale, as being those which are most directly needed in the present state of the science, I feel that I can acquiesce with very good grace in the substitution for them of a set of valuable detailed experiments upon models, conducted under such superintendence as we may depend upon these receiving from Mr. Froude. We have far too much to learn, not to be glad of any carefully-conducted experiments on the resistance of ship-shaped forms, whatever may be the absolute dimensions of the subjects of experiment…"[3]

A less gracious and more outspoken critic of Froude's plan was renowned Scottish naval engineer, Scott Russell, who not only founded the Institution of Naval Architects[4] but had also built the SS *Great Eastern* with I.K. Brunel a decade before. Russell was disappointed that the results of model tests he himself had run resulted in imprecise findings when applied to full-sized ships. Convinced by this personal experience that model tests would not only be a waste of time but lead

to inaccuracies, he was adamant in his opposition. He did allow that such model testing might point indirectly to the proper direction for full-sized testing, however.

At a meeting of the Committee in 1870, he said, "I do not announce to you that
experiments on very little models are very safe *data* for experiments on large ships,
but I wish you to possess all the information I do upon the subject, in order that you
may not be disappointed when Mr. Froude brings you the results of his experiments
on little models in a canal, and in order that you may not expect from those to obtain
data which will enable you to go right at the construction of large ships for the open
sea. The reason why I prepare you not to expect such reliable results from little
models is this: that I myself have taken the trouble to make a series of experiments
on 120 small models... Indeed the most interesting fact I ascertained was, that the
results on a large scale, were precisely the contrary to the results on a small scale.
But it was very interesting to me, and the most agreeable period of my life was that
romantic period of about two years in which I was mainly occupied with the amusement of making pretty little experiments on a small scale."[5]

Froude was undeterred and pointed out that neither of the formal objections to
the use of models would hold true in the case of a model scaled properly at a minimum of 6 feet in length. In response to questions from the Committee, he said, "...
I see that the feeling of the meeting is very much against experiments with models,
but I must say that my own experience leads me to judge quite differently. I think
the reason why experiments with models have hitherto been found to be a failure,
and have misled those who have made them, as to the effect to be expected with
regard to a full-sized ship, is, that attention has not been paid to the relation which
should subsist between the speed at which the model is moved, and the speed at
which the ship is moved." He concluded his remarks, "... I believe we have still a
great deal to learn, both practically and theoretically, and though I cannot set my
reputation and credit against that of the various persons who have addressed you
to-day, all I can say is that I believe, myself, I shall get a great deal of useful information from the experiments which I propose to make."[6]

Froude based his assessment of the possibilities of model testing on several trials
he'd previously run in Dartmouth Creek. These tests had been made on 3-, 6-, and
12-foot scale models of two hull designs. The "Raven" was constructed in keeping
with Scott Russell's "waveline" theory and had a sharp bow (Fig. 6.1). The "Swan"
had a blunter prow (Fig. 6.2). During the tests, Froude noted that models run at
speeds that were proportional to the square of their length generated wave patterns
that were virtually identical.[7] He also noted that the Swan's blunter bow gave her the
advantage. Froude was certain that more scientific tests would result in better data,
data that could be used by the Royal Navy in the design of her ships. It was with this
work with the Raven and Swan fueling his belief that he'd approached the Admiralty
for funding.

Froude submitted an application for a model tank with the support of Sir Edward
Reed. His application for an experiment tank was approved in February 1870. He
was given £2,000 to construct and run the tank for 2 years. The first model was
tested on March 3, 1872.[8] Work actually went on there until the lease ran out 14
years later. At that time the works were dismantled, and operations were transferred

Fig. 6.1 The Raven. The Raven was constructed in keeping with Scott Russell's waveline theory

Fig. 6.2 The Swan. The Swan had a blunter bow

to a facility in Haslar where a flask of water from the Torquay tank was added to the new basin. The initial monies given to Froude included wages for some assistants and his son, Robert Froude, who carried on the work after William Froude's death in 1879. William Froude himself was never paid for his time, having volunteered for this important work. His work was highly valued, however and upon his death the Admiralty sent a message to his son that concluded with, "My Lords desire to convey to you, and other members of the family, the expression of their most sincere sympathy at the irreparable loss which you have sustained—a loss which cannot be looked upon as other than a national one."[9]

Froude's Early Work

Froude had earned the respect of the naval community with his early work on the rolling of ships. This was a great problem of the time because the factors at work in the rolling of ships were poorly understood. As a result, there was no way to plan for a ship's reaction to waves and the performance of ships suffered accordingly in terms of passenger comfort and overall stability and seaworthiness. I.K. Brunel approached Froude before Brunel's death in 1859, and asked him to give the matter consideration as it applied to the design and construction of the *Great Eastern*.

Froude approached the topic with his signature meticulous work. He wrote in a seminal paper presented at the Second Session of the Institution of Naval Architects on March 1, 1861, "The most observable feature in the actual movement of a ship when rolling, and that which had always appeared to me to be specially characteristic of the dynamical laws to which it would be necessary to refer them, is the gradual accumulation of angle during several successive rolls; the cumulative action thus growing up into a maximum, and then dying out by very similar gradations, until the ship becomes for a moment steady, when a nearly similar series of excursions commences and is reproduced: while in reference to the momentary pause, or cessation of motion, it has seemed to me clear that it occurs, not because the waves themselves cease, or cease to act, but because the last oscillation has died out at a moment when the ship and the waves have come to occupy, relatively, a position of momentary equilibrium."[10]

In his paper he also stated that as a result of his work and others he was convinced that waves had a cumulative effect and that, "this aspect of the question is so closely analogous to what happens when any oscillating body, such as a pendulum, is subjected to a series of impulses, partially synchronous with its own excursions, that it had always seemed to me probable that the laws which govern the latter class of phenomena, would be found, *mutatis mutandis,* applicable to the elucidation for the former also; and in attempting to investigate regularly on this line of thought the dynamical relations of a ship, and of the waves on which she floats, it turned out that the solution a less difficult than had been expected, and that its fundamental results, at least, could be arrived at with considerable completeness and closeness of approximation."[11]

The question Froude set out to answer was, "what is the *position of momentary equilibrium* for a body floating on a wave, and what accelerating force towards that position will the body experience in terms of her momentary deviation from it?" Froude realized that the action of a water particle is governed by gravity when the water is still. When the water is in motion, the particle is not only subject to gravity but to accelerating forces as well. The net result of all these forces is an incline that can be calculated if the direction and magnitude of the forces is known.[12]

To test his theory in the research facility he had housed in the basement of his home, Froude reported, "a float was formed of cork, somewhat like a small lifebuoy, about four inches in diameter; a mast was planted obliquely in one side of it, with its apex perpendicular over the centre of the float; a small plum-bob was suspended from this, having its centre at the level of the center of buoyancy of the float, and occupying, when in still water, the centre of the ring. When this was set afloat in a trough, fitted with apparatus for generating waves, while the plane of its flotation followed the slope of the waves, the plum-bob remained, nevertheless, so completely central, that to an eye resting on it, it was difficult to believe that the surface was really disturbed by waves, though on watching the sides of the trough it was plain that the wave slope ranged up to 15 deg. or 20 deg.; the plumb-line, at the same time, deviating to the same extent from the perpendicular."[13]

An important consideration in the behavior of a ship is the ship's "period." In his paper, Froude wrote, "It follows, farther, that when the ship at any moment deviates from this position, the effort by which she endeavours to conform herself to it depends on the momentary angle of deviation, in the same manner as her effort to assume an upright position, when forcibly inclined in still water, depends on the angle of inclination. Hence her stability, i.e. her effort to become vertical

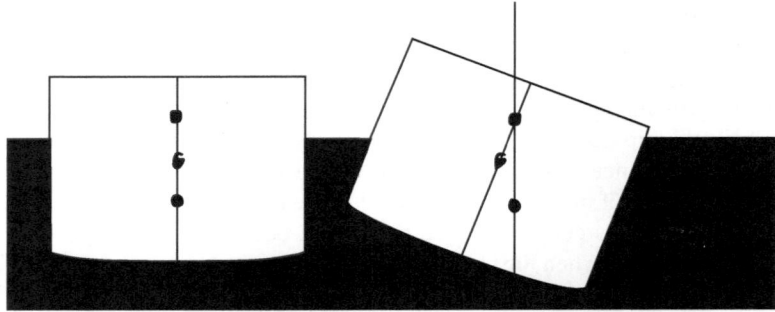

Fig. 6.3 Metacenter. The metacenter changes to maintain a position directly above the center of buoyancy

in still water, measure her effort to become normal to the waves in undulating water: and hence just as when the ship floats in still water, this measure of the effort, changing with her changes of inclination, combined with the measure of her 'moment of inertia,' serves to determine her period of oscillation; so when she floats in waves, the effort, similarly measured, and changing not only with her own changes of inclination, but also with those of the traveling wave surface, serves to determine the successive changes of position which she will then experience."[14]

Froude found that the cork would work to maintain a position at right angles to the surface of the wave, whether the water was still or a wave was inclined. The current theory of the *metacentre* explains this phenomenon. The metacentre is located directly above the *center of buoyancy* of a ship. When a ship is in still water, the metacentre, center of gravity, and center of buoyancy are in a direct vertical line. When a ship is in waves, the metacenter will change to maintain its position directly above the center of buoyancy as that shifts laterally with the motion of the ship in the wave. The shifts are automatic as the ship strives to maintain the metacenter above the center of buoyancy (Fig. 6.3).

"We see, then," Froude wrote, "that though the effort of stability of any given ship depends primarily on her mass, and the position of her metacenter and her center of gravity, the rate at which she will acquire or lose velocity under given circumstances of inclination and angular velocity, and the position she will assume at any period, may be wholly express in terms of her "periodic time" *T*."[15]

Froude undertook his work on the rolling of ships with the goal of bringing science to a vexing problem. He was unhappy that current practice resulted in outcomes, "so that when a new ship is sent to sea, her constructor has to watch her behaviour in a sea-way with as anxious and uncertain an eye as if she were an animal he had bread and was rearing, and hoped would turn out well: not a work which he had himself completed, and whose performance he could predict, in virtue of the principles he had acted on in its design."[16]

In Appendix I to his 1861 paper, Froude laid out a number of conclusions and went into further detail on the calculation and application of (*T*), the number of seconds taken for a ship to make a single oscillation in still water and (T^1), the number of seconds between the wave in "passing from hollow to crest, or crest to hollow."[17]

The first conclusion was that, "all ships having the same 'periodic time,' or period of natural roll, when artificially put in motion in still water, will go through the same series of movements when subjected to the same series of waves, whether this stability in still water (one of the conditions which governs the periodic time) be due to breadth of beam, or to deeply stowed ballast, or to any such peculiarity of form as is in practical use.

The second was that, 'the condition which develops the largest angles of rolling is, equality in the periodic times of the ship and of the waves…' This is because the period of the waves would amplify the effects of the period of the ship, resulting in larger angles than would normally occur."

The third was that, "the ship will fare the best which, *cateris paribus*, has the slowest periodic time." This was because the ship would be less reactive to the period of the waves.

And the fourth was that, "there are two, and only two, methods of giving a slow period to a ship.

(a) By increasing her 'moment of inertia,' as by removing her weights as far as possible from her center of gravity; an arrangement which for the most part can only be accomplished to a limited extent.
(b) By diminishing her stability under canvas. This can always be accomplished in the construction of a ship, and generally in her stowage, to any degree consistent with her performance of her regular duties, by simply raising her weights."

With his work and his formula for the calculation of (T) and (T^1), Froude had not only thoroughly examined and explained the rolling of ships, he had supplied shipbuilders with a scientific way to calculate the future performance of their ships.[18]

The Admiralty Experiment Works at Torquay

Plans for the Works

Froude's work on the rolling of ships had earned him a reputation for meticulous work and innovative thinking long before he approached the Admirality for funding for his Experiment Works. Once the funding had been secured for a facility, he applied that same attention to detail to the specifics of his plans for his basin. The plans called for water space that was exclusively devoted to the purpose of scale model testing so that it would be free from currents produced by other vessels. He also wanted dedicated water space so that a uniform water level, free from weeds and other obstructions, could be maintained. He wanted the water space housed so that experiments could be run without interference from inclement weather (Fig. 6.4).

Fig. 6.4 Image of basin. The Torquay Experiment Tank, 1871. *Source*: *The Papers of William Froude, M.A., LL.D., F.R.S. 1810–1879*

He further specified that the waterway itself must be of sufficient width and depth so that additional resistance would not be caused by "constriction of the fluid." The basin also needed to be long enough to give the model ample time to attain the optimal speed and to maintain that speed for a duration that allowed the test to be run. Froude estimated a maximum width of about 10 feet and a length of about 250 feet would be sufficient.[19]

Froude's careful planning extended to the composition and dimensions of the models. He considered a 6-foot model to be sufficiently large although he estimated that the model size could go up to 12 feet. He also concluded that the models would not need to be stored once tested as long as the dimensions were saved. New models could be built to existing dimensions if needed again in the future.[20]

He further required the models be made of a material that was easy to cut. Froude suggested stearine, a type of wax that was firm enough for the purpose. But the entire model was not to be made of stearine; Froude proposed "an internal shell of wood which will carry the fittings for ballast, towing and handling…" The stearine would be "roughly 'melted on' in a clay or other mould to a sufficient thickness to allow for subsequent shaping."[21]

Once cut to their initial shape, the models could be "altered to others by a revolving cutter mounted on a traversing frame… the path of the cutter instead of following the surface of a solid pattern will follow on a drawing the successive water-lines, the cutter being set successively at the levels corresponding to the water-lines."[22] (Figs. 6.5, 6.6, 6.7, and 6.8).

MACHINE FOR SHAPING MODELS USED IN
EXPERIMENTS ON FORMS OF SHIPS

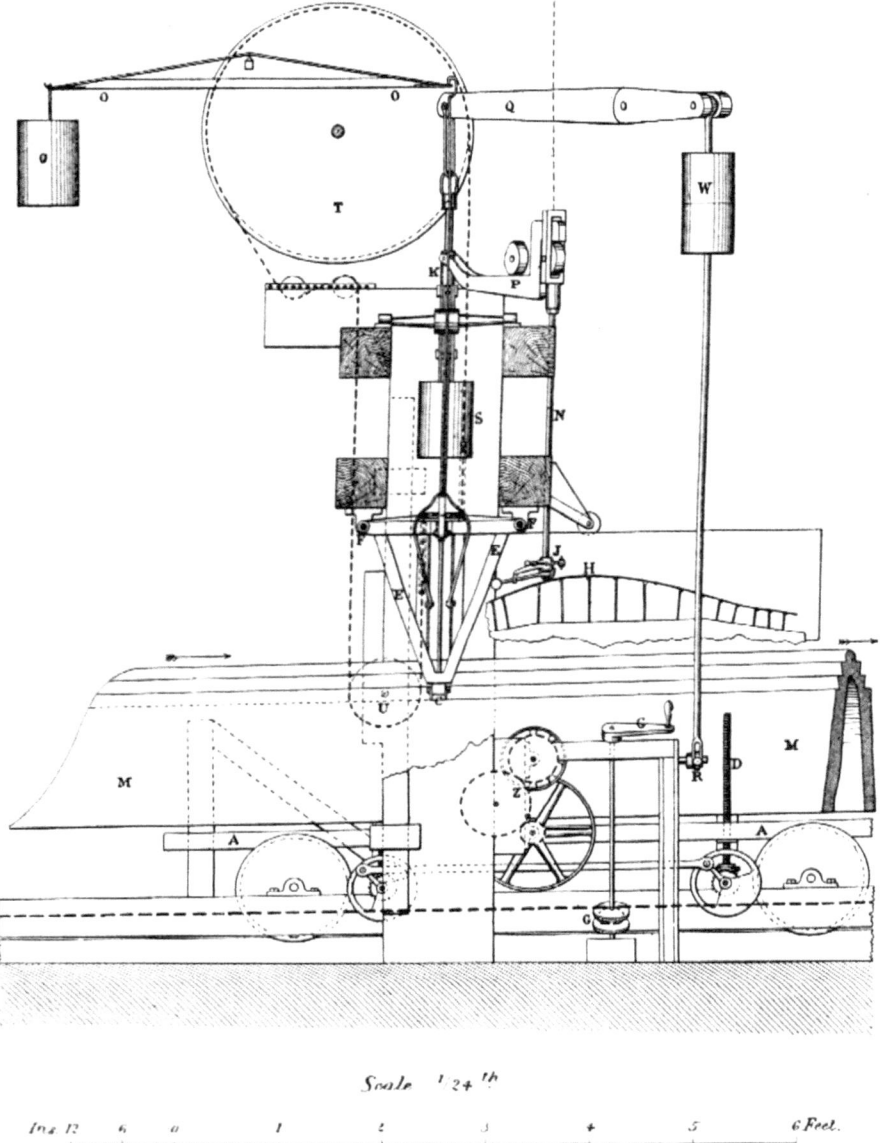

Fig. 6.5 Image of cutters. Side view of machine used for shaping the models. *Source*: "Description of a Machine for Shaping the Models Used in Experiments on Forms of Ships" by William Froude, 1873

DESCRIPTION OF A MACHINE FOR
SHAPING THE MODELS

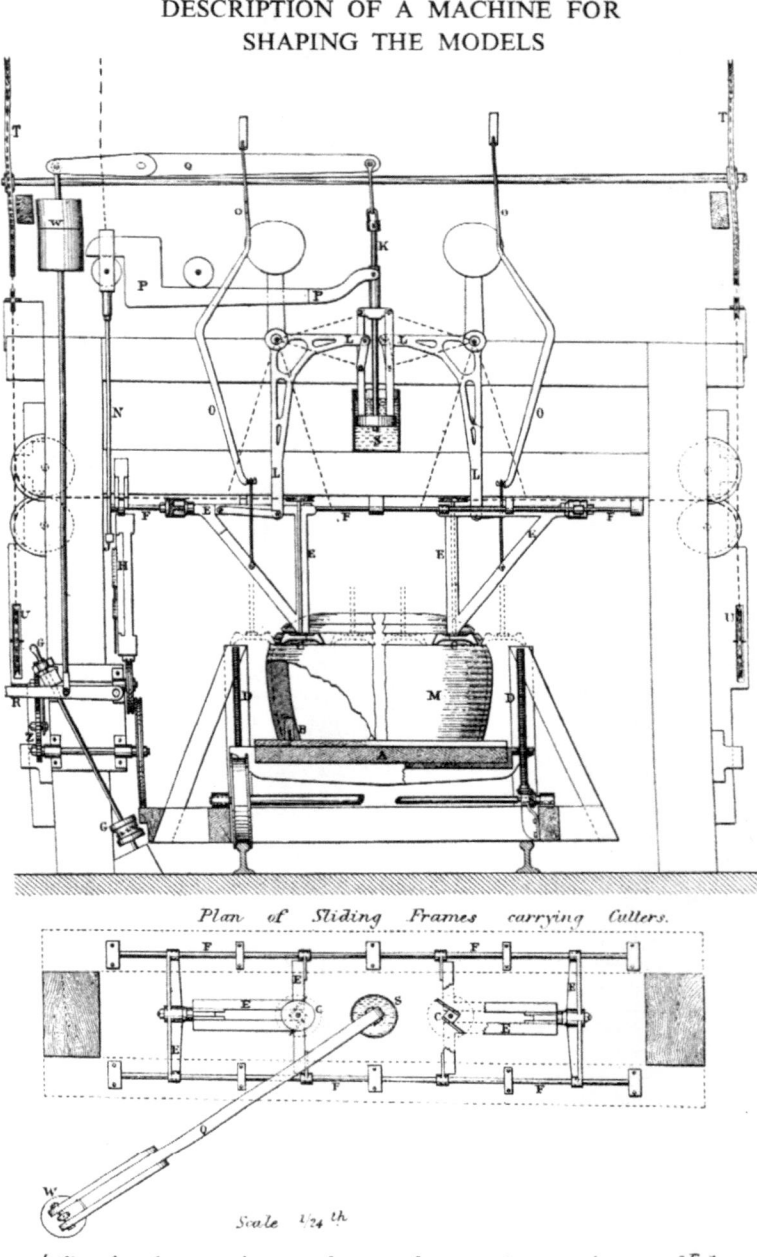

Fig. 6.6 Image of cutters. End view of machine used for shaping the models. *Source*: "Description of a Machine for Shaping the Models Used in Experiments on Forms of Ships" by William Froude, 1873

USED IN EXPERIMENTS ON FORMS OF SHIPS

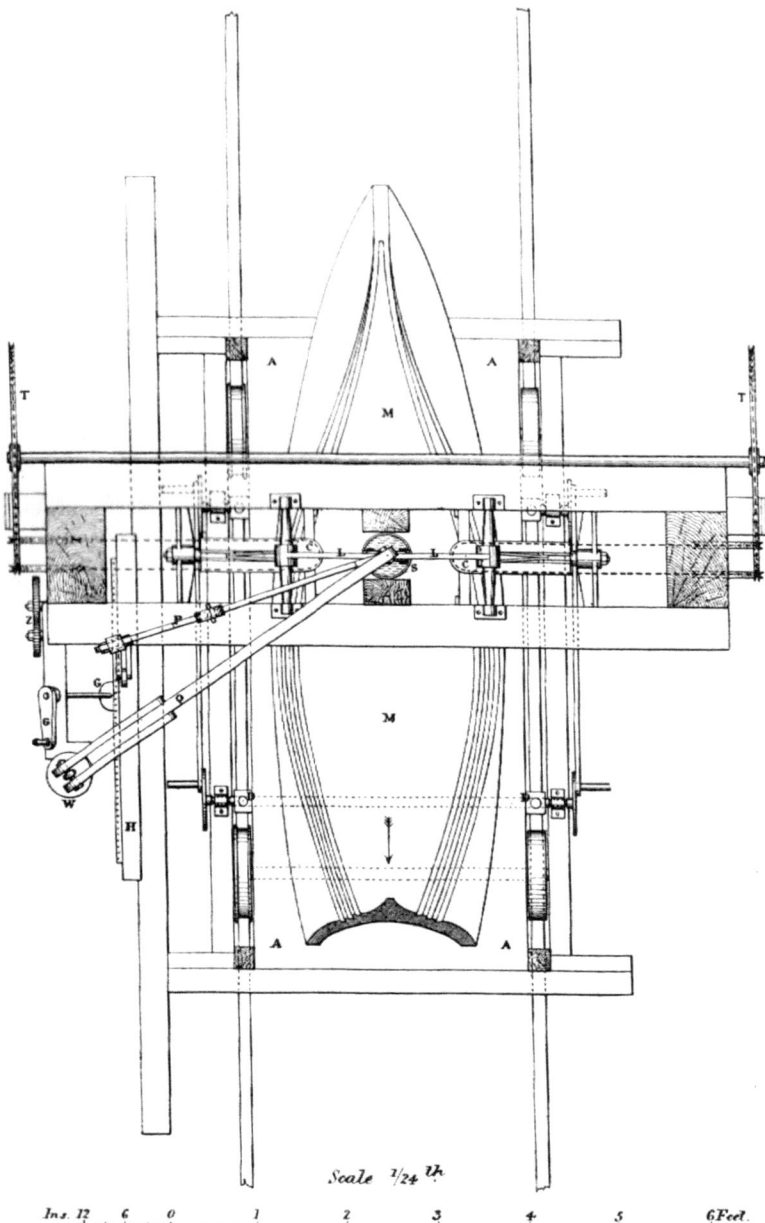

Fig. 6.7 Image of cutters. Plan of machine used for shaping the models. *Source*: "Description of a Machine for Shaping the Models Used in Experiments on Forms of Ships" by William Froude, 1873

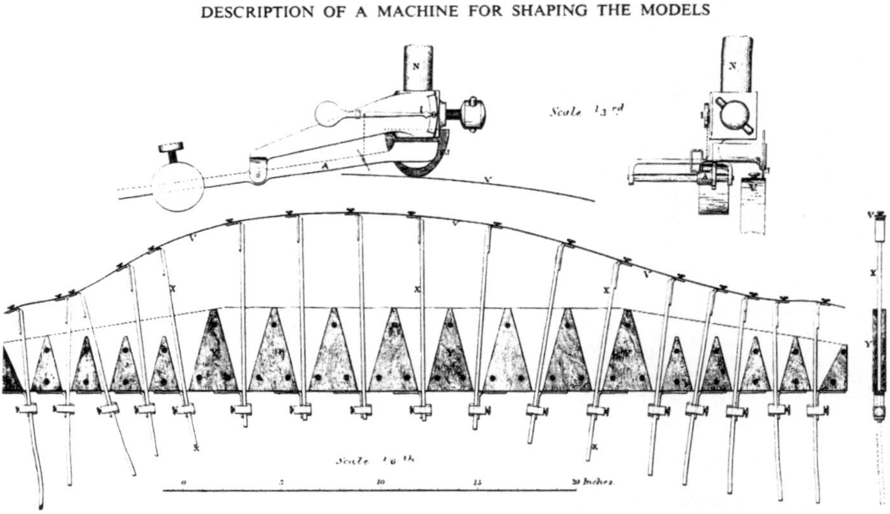

DESCRIPTION OF A MACHINE FOR SHAPING THE MODELS

Fig. 6.8 Image of cutters. Template. *Source*: "Description of a Machine for Shaping the Models Used in Experiments on Forms of Ships" by William Froude, 1873

The next thing Froude would need was an apparatus to measure the relationship between velocity and resistance on the various models. "The prima facie simplest way of doing this is that which has been generally adopted, namely that of applying a definite suspended weight acting through a delicate tow line as the tractive force, which force is thus in each experiment regarded as the measure of the resistance, and noting the time taken by the model to traverse certain distances."[23]

Froude knew that a weight attached to one end of a rope attached to the model would do the job of pulling the model through the water, yet he worried that since the results would not be self-recorded, they must be observed and noted with the possibility for observer error. He also had concerns about any elasticity in the tow-line that might interfere with the results. More worrisome still to Froude was what he referred to as a "still graver difficulty," namely, the near impossibility of guiding the boat on an exact straight path. Froude's experience told him that "no model can be trusted to follow exactly the direction of the tow line, indeed some forms, in which the tendency to sheer would be least suspected, I have found to possess it to an astonishing degree."[24]

The problem of how to tow the model in a straight line was a big one. Any deviation from a straight line would add to the time it took the model to cover the distance and would be added to the resistance of the model. In reality it might not be due to the design of the ship and if it were, it would not be possible to determine which part of the design was causing the problem. In his search for a solution, Froude considered attaching guide wires to the models but feared that would introduce other areas of error. He ultimately decided upon what we now call a *towing carriage* as the optimal method for propelling the models. Froude wrote, "If however,

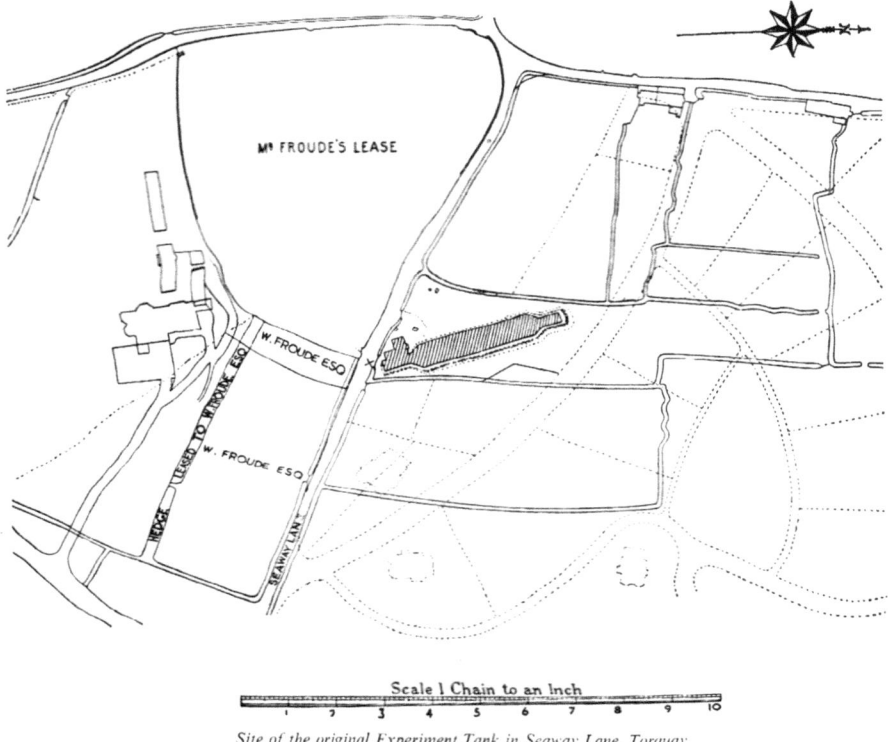

Site of the original Experiment Tank in Seaway Lane, Torquay

Fig. 6.9 Torquay plot plan. The Torquay Basin and land. *Source: The Papers of William Froude, M.A., LL.D., F.R.S. 1810–1879*

we substitute for the guiding wire a light railway with a truck travelling on it we may attach the model to the truck through the medium of a dynamometer, affixing the tow line to the truck instead of to the model, and the dynamometer while it guides the model inexorably will indicate purely and solely the resistance experienced by the model in traversing the water. The rotation of the truck wheels will afford an exact measure of the distance as it is travelled by the model; and a 'time apparatus' such as I used in my former experiments, being added, we have all the necessary elements, which when duly combined in one apparatus will record by lines traced on a travelling sheet of paper the velocity and resistance of the model at each point of its course."[25]

With a towing carriage, Froude would no longer need the weight and towline. Instead he would use a motor with a governor that would maintain a uniform speed. "No doubt all this entails an apparatus of somewhat costly construction," Froude wrote, "but after giving the subject much trial and consideration I am confirmed in the opinion that it is the only sure way of obtaining a suitably accurate result."[26] The use of the towing carriage also eliminated the need for an independent observer to record the progress of the model. With resistance thus precisely measured as the model progressed, Froude was satisfied that his results would be properly recorded. In hindsight it is possible to state that Froude was correct in his conclusions about the effectiveness of the towing carriage as he described for precise measurement. Towing carriages with recording equipment are still in use in model basins around the world to this day.

The Torquay experiment tank was located in a field near Froude's home (Fig. 6.9). The facility was always designed to be temporary in nature because the lease called for the land to be returned to its original condition at the term. The Works consisted of a canal, an office, and a workshop located in a wood building with a brick boiler-house nearby. The waterway was the focal point of the facility. At 278 feet long, 36 feet wide at the surface, and 10 feet deep, it was formed by a combination of excavating dirt and building up the sides with the soil removed during the digging to form an embankment. To keep the embankment from leaking, the sides were puddled and surfaced with asphalt. When the scale was considered, the 36-foot width was the equivalent of a channel of about 2,000 feet across. It was also equivalent to 33 fathoms in depth when modeling a frigate to scale. In keeping with Froude's requirement that the water be unconstricted, this great width ensured that any waves created would dissipate without interfering with the tests being run. It was part of the reason Froude was confident his tests would be accurate. Another was the fact that there was a "beach" at the far end of the tank that permitted the dissipation of waves created by models as they were towed through the water.[27]

The Equipment

The towing carriage Froude specified would be used to move the models through the water without creating an additional disturbance. The carriage would be entirely out of the water and as a result would generate no waves or interference of its own. This carriage would be far more sophisticated that the crude use of ropes with weights attached, wherein the non-weighted end of the rope was attached to the model, the weight was dropped into a large hole, and the model lunged forward. The towing carriage met Froude's desire for something more scientific for his new experiment tank, something that would pull the models at a steady and measurable rate. It utilized a 3-foot, 3-inch gauge railway for the towing equipment and screw dynamometer, "carried by the roof principals and extended the full length of the waterway. This arrangement enabled the carriages to be of light construction..." The resistance carriage and screw carriage could be run individually or connected to one another. They were towed by an endless wire wound around a barrel driven by "a stationary steam hauling engine at the starting end." Another small carriage could be attached behind the resistance carriage so observations of wave formation at a model's stern could be made.[28]

The steam engine, designed by Froude, could tow at a speed of up to 1,000 foot per minute. If greater speeds were needed, Froude would go "old style," and attach

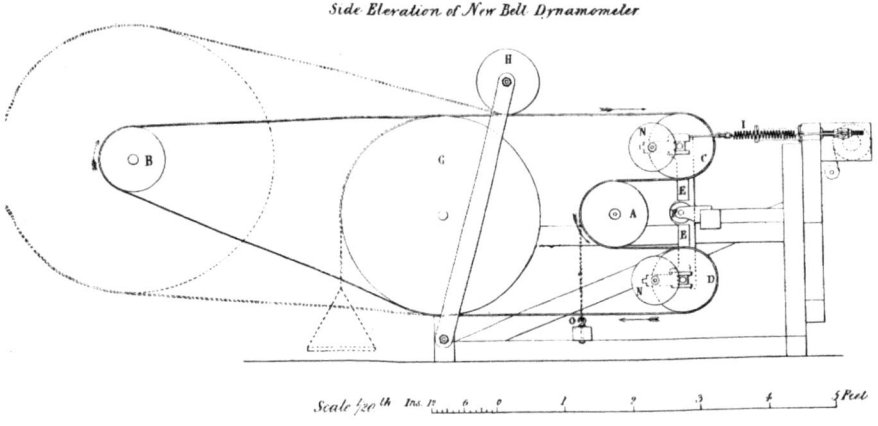

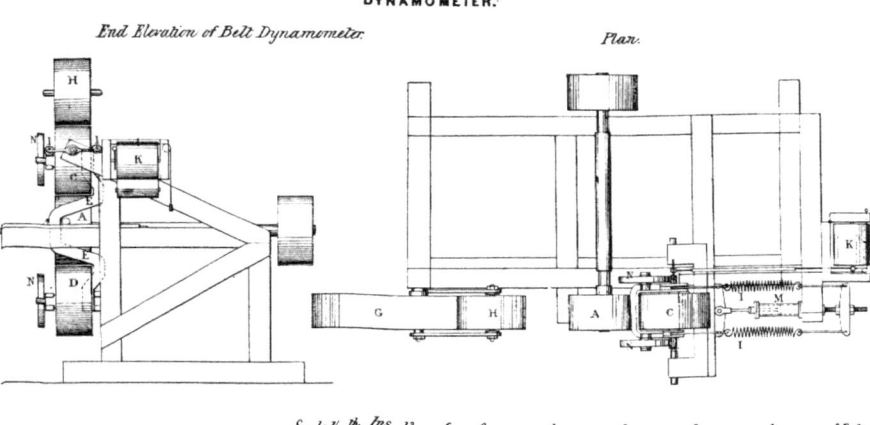

Fig. 6.10 Image of belt dynamometer. The belt dynamometer was designed to record the performance of any power consuming machine which is or can be driven by a belt. *Source*: "On a New Dynamometer and Friction Break" by William Froude, 1858

the carriage to a rope with a weight at the other end. The weight would then be dropped into a pit outside the building. With the towing apparatus maintaining a steady speed above, rather than in the water, Froude had what he needed to propel the models without introducing additional motion in the water.

All of this effort was in the service of measuring the amount of friction, resistance, the model encountered as it moved through the water. Would a long hull have less drag? Would a wider hull make a difference? Did the amount of power or the size of the screw propeller alter the performance significantly? The proper calculation of the resistance was vital.

Froude designed a variation on a belt dynamometer for the job (Figs. 6.10 and 6.11). This piece of equipment was attached to the towing carriage so that when

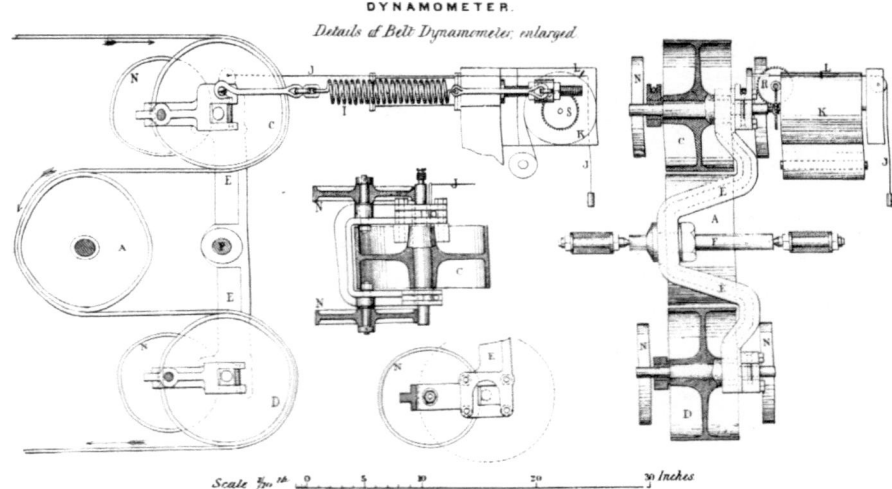

Fig. 6.11 Image of belt dynamometer. Details of the belt dynamometer. *Source*: "On a New Dynamometer and Friction Break" by William Froude, 1858

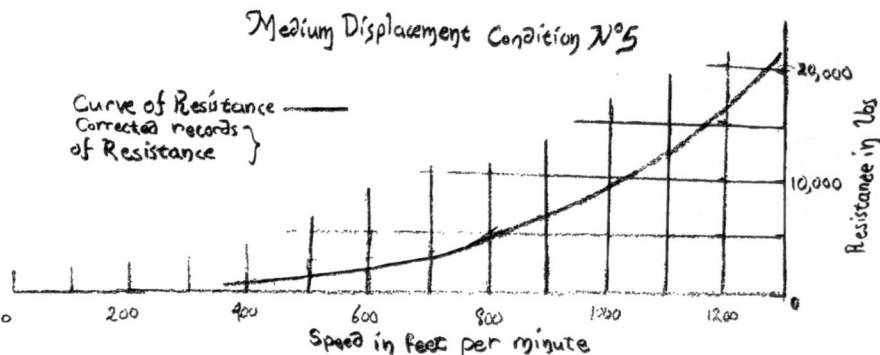

Fig. 6.12 Resistance curves. The effects of changes in design on the resistance of the model

the model was towed, it made two lines on a long piece of paper. One line was straight and documented the total time the vessel was in motion. The other line was made by a pencil at right angles to the paper and attached to a spring mechanism that permitted the pencil to move up and down on the page and draw a curved line. The distance between the curved line and the straight line showed the effect of friction, the resistance that particular ship form encountered when moving at that particular speed. To make it easier to keep track of the amount of time that had passed, hash marks were made on the paper to document each half-minute. The curves were compared to one another to see the effects of speed and shape on resistance. By varying the speed and/or the form, Froude could observe the effect variations in design had on the model and extrapolate to their effect on full-sized ships[29] (Fig. 6.12).

The Models

In addition to meeting Froude's desire for ship form models that could be easily cut, the models themselves were works of art. Each had to be manufactured to exact specifications so that test results would be comparable. To ensure a smooth, defect-free surface, the wooden cores were coated with hard paraffin wax. A small amount of beeswax was added to guarantee there would be no air holes in the finished surface. Froude was delighted with his selection of materials. He wrote in a paper published in 1873, "The material best suited for the purpose would clearly be one, which, though hard enough, would cut freely and smoothly in any direction, without requiring much power and without blunting the tools; a material fusible at a low temperature, so as to be easily cast approximately to the shape required; of not great specific gravity; impenetrable to water without the application of any artificial coating; and, above all, capable of being melted up again repeatedly to form new models. Almost the first material that suggested itself—hard paraffin—was found to fulfil every one of these conditions."[30]

There had never been models constructed to the specifications Froude had in mind. He needed a new type of machinery for the shaping of the models. Froude designed and built the machine himself. The resulting machinery and process were so exact that "it would certainly be considered there was something wrong in the model, if when properly loaded the addition of only ½ lb. weight would not make it deviate sensibly from the flotation line calculated for it."[31]

Initial Tests

The initial Experiment Tank tests were made on plain wooden planks. Froude reported his findings before the British Association for the Advancement of Science at Brighton in August 1872. The object of the experiments was to discover the conditions of the resistance to passage through the water caused to models or ships by the friction of the water against the sides. The planks were shaped to various scales and towed through the water at varying speeds as the results were recorded. The planks were then treated with a number of varnishes that included shellac, tallow, and glue, and towed again.[132] The results were compared to see what, if any, difference the various surfaces might make. The smoother surfaces proved to have less resistance.

Paraffin scale models were then created and towed through the water. The results from these tests were compared to the tests with the planks, to see what effect the shape of the model hulls had on the friction, or resistance, experienced by the model. They showed that the different shapes had different resistance and that the resistance was not only present, but measurable. All Froude needed now was a way to prove beyond doubt that his 6-foot paraffin scale models supplied results that were applicable to full-sized ships.

Froude went on to make a number of important discoveries using models of various forms with his towing carriage and dynamometer. He tested the resistance they encountered at differing speeds and in his 1877 report read at the Eighteenth session of the Institution of Naval Architects, March 23, 1877, he gave the first formal explanation of his "system of experiment," which included his rationale for testing and comparing many possible forms. "That system of experiments involves the construction of models of various forms (they are really fair-sized boats of from 10 to 25 feet in length) and of testing, by a dynamometer the resistances they experienced when running at various assigned appropriate speeds. The system may be described as that of determining the scale of resistance of a model of any given form, and from that the resistance of a ship of any given form, rather than as that of searching for the best form; and this method was preferred as the more general, and because the form which is best adapted to any given circumstances comes out incidentally from a comparison of the various results."[33]

Froude went on to explain the vital function of the dynamometer in the process. "We drive each model through the water at the successive assigned appropriate speeds by an extremely sensitive dynamometrical apparatus, which gives us in every case an accurate automatic record of the model's resistance, as well as a record of the speed."[34] The fact that the recording was automatic and accurate was a major milestone in the collection of data of this sensitive nature.

He also included a description of the arrival of a resistance curve of each model. "We thus obtain for each model a series of speeds, and the corresponding resistances; and to render these results as intelligible as possible, we represent them graphically in each case in a form which we call the 'curve of the resistance' for the particular model."[35]

This introduction of his method also included a brief discussion of the purpose of the curve. "This curve, whatever be its features, expresses for the model of that particular form, what is in fact and apart from all theory, the law of its resistance in terms of its speed; and what we have to do is if possible to find a rational interpretation of the law."[36]

Froude ended the introductory portion of his remarks with a brief discussion and definition of resistance. The fact that the discussion was brief does not detract from their importance since Froude was the very first to design and describe the method that would be used going forward.

He pointed out that an earlier belief about the movement of ships through the water—that they plowed their way forward—was definitively proven to be false by his work. "The old idea that the resistance of a ship consists essentially of the force employed in driving the water out of her way, and closing it up behind her, or, as it has sometimes been expressed, in excavating a channel through the track of water which she traverses, this old idea has ceased to be tenable as a real proposition, though *prima facie* we know that it was an extremely natural one."[37] This was an important point because until it was acknowledged that the resistance encountered from a ship did not come from "making a channel" as it progressed, there was no way to discuss the origin of resistance and practical ways to reduce that resistance.

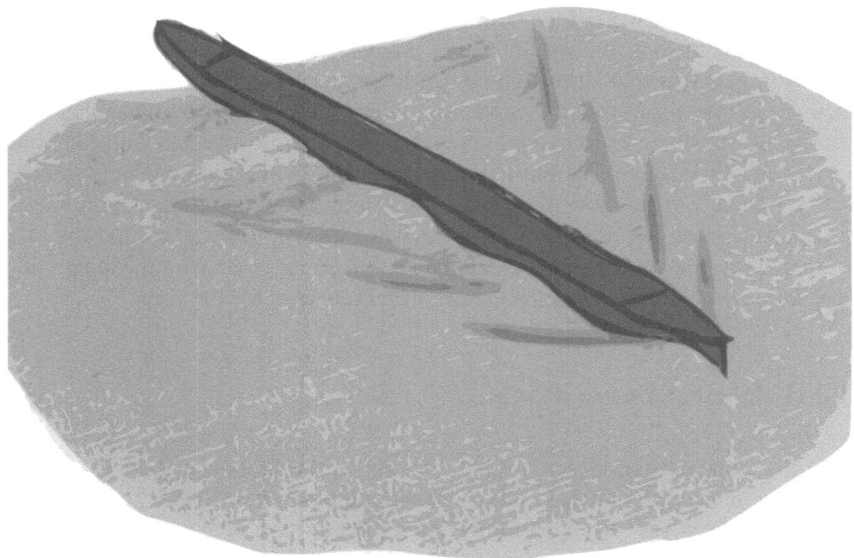

Fig. 6.13 Transverse waves. Waves which appear independently at the stern of a ship

Froude wrote, "Thus we divide the forces represented by the curve of resistance into two elements—one 'skin resistance' the other, which only comes into existence as the speed is increased, and which we may term 'residuary resistance'."[38] As you've probably concluded, skin resistance was due to the wetted surface of the ship and was nearly identical to the total resistance encountered by the vessel at slow speeds.[39] The residuary resistance was trickier.

Residuary resistance, Froude explained, comprised the greatest form of resistance and was made of three components. The first form of resistance arose from the energy that went into creating the bow waves that diverged from the ship, never touching the ship again once they were generated and dissipating as they traveled away from the vessel.[40] The more important forms of residuary resistance came from the transverse waves, "the crests of which remain in contact with the ship's side, and thirdly the terminal wave, which appears independently at the stern of the ship."[41] (Fig. 6.13). The terminal waves were formed as a result of water that had travelled the length of the ship beneath the water and rose to the level of the water line as it passed under the farthest portion of the ship.[42]

Froude concluded his introduction by writing, "The term 'wave-making resistance' represents then the excess of resistance beyond that due to surface friction, and that excess we know to be chiefly due to this formation of waves by the ship."[43]

As a result of the work done with his system of experiment, Froude was able to reach two conclusions that would impact the design of vessels from that point forward: The first was that ships that tended to produce longer waves also tended to reach greater speeds before the waves were generated. The second was that such a ship would encounter less wave-making resistance at most speeds.[44] This led Froude to conclude that "this principle is the explanation of the extreme importance of having at least a certain length of form in a ship intended to attain a certain speed; for it is necessary, in order to avoid great wave-making resistance, that the 'wave features,' as we may term them, should be long in comparison with the length of the wave which would naturally travel at the speed intended for the ship."[45]

The HMS *Greyhound* Trials

When the Admiralty proposed a series of tests on the full-sized HMS *Greyhound*, Froude recognized his chance. Given the strength of opinion that full-sized tests were a must, he wisely arranged for the Admiralty to have him appointed one of the two people conducting the full-sized mile trials of the HMS *Greyhound*. The purpose of the trials was to determine "the resistance experienced by a ship of considerable size and of known form and dimensions, when moved through smooth water at various speeds."[46] To accomplish this, Froude had the actual *Greyhound* towed by a larger ship by use of a 45-foot boom. This boom allowed the *Greyhound* to be towed without being impacted by the wake of the towing vessel (Fig. 6.14). The weather was good and the tests were run on the 878-ton ship (without her masts) at varying trims and displacements.

Among other things, Froude was able to determine that about 58 percent of the power was being wasted in the friction of engine and screw and in the "detrimental reaction of the propeller on the stream lines of the water closing in on the stern of the vessel." [47] This finding would lead to further investigation of propellers. He also came up with a series of resistance curves for the actual, full-sized HMS *Greyhound*.

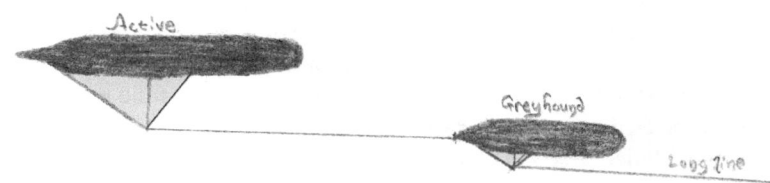

Fig. 6.14 Greyhound trials. The trials of the HMS *Greyhound* proved the validity of scale model testing

Froude than ran tests on his scale model of the *Greyhound* in the Torquay Experiment Tank. He said, "In order to test the 'scale of comparison' which has been propounded by me as furnishing a true method of inferring the resistance of a ship from those of a model of the ship, a model of the *Greyhound* 1/16 full size was made, and its resistances at various speeds determined under each of the different conditions of displacement and trim to which the ship herself was subjected. This was done in the experiment-tank, and with the apparatus constructed by me for the experiments I am now carrying on with models for the Admiralty."[48]

He discovered that when the dynamometer results of the full-sized trial were compared to the scale trials of a model with a hull that was not completely smooth (as was the case with the full-sized *Greyhound* due to chemical reactions on the surface of her copper plated hull) the results were a match and "...conclusively verified], the law of comparison between ships and models, the discrepancy which it presents being only such as might arise in comparing the performances of any given ship under two different conditions of skin."[49]

This indicated that what Froude would have expected to be the case with the actual *Greyhound* given the results from the scale model tests did indeed hold. His scale model testing had given results that were verifiable and accurate.

With the completion of the *Greyhound* trials and the successful comparison of the mile test results to the results of the scale model trials, Froude's faith in the accuracy of scale-model testing was affirmed. "This justifies the reliance I have placed on the method of investigating the effects of variation of form by trials with varied models—a method which, if trustworthy, is equally serviceable for testing abstract formulae, or for feeling the way towards perfection by a strictly inductive process," he wrote in a paper that proved to be definitive.[50]

Froude's work established model testing as a valid and vital precursor to the design of ships of the fleet. By 1894 all British warships were constructed using input from the results of model testing. The results of model testing work affirmed the faith Froude had placed in the value of his earlier work—work that Scott Russell has once derisively compared to as *the amusement of making pretty little experiments on a small scale.*

Conclusion

Froude's goal had been to use scale model tests to properly estimate the amount of resistance experienced by a full-sized ship as it moved through the water. This was a vexing problem of his time, because the dimensions of a battleship were not calculated using any sort of consistent rationale. Froude wanted to help take this important problem from an art to a science, and to do this he used a dynamometer to chart the progress of a scale model through the canal he constructed at Torquay. This would be done with thousands of models over the 14 years the experiments took place in his facility.

The tests would vary to simulate different dimensions of the ships, to see if a wider, shorter ship did better than a thinner, longer ship, for example. Froude was convinced earlier problems with model test results would be resolved by the use of scale models. His belief was vindicated with the mile trials of the actual HMS *Greyhound.* For this test, the full-sized *Greyhound* was towed by another ship. Different conditions of trim were introduced and the results were measured and compared to the model tests. When the tests showed the same results, it laid all doubts to rest and model testing was firmly established as the precursor to ship design for ships of the British fleet.

Hydrodynamicists respect William Froude's contributions to hydrodynamic theory and practice to this day. His work not only established the validity and usefulness of model testing, but resulted in several theories that have stood the test of time:

Froude's Comparison is predicated on the fact that observations made on a scale model are valid indicators of the observations that will be made on the full-sized ship under the same conditions. This is a seminal concept that made it possible for naval architects to "try out" a number of theories on models, before any expenditure was made on actual ships of the line. It allowed science to be applied to ship construction and in the process ensured that the finished ship would perform as anticipated. This was a huge step forward from the rules of thumb and application of personal experience that had ruled the design of ships until Froude's time.

In "The Speed and Power of Ships," first published in 1910, Rear Admiral David Watson Taylor, constructor and head of the Experimental Model Basin at the Navy Yard in Washington, DC, devoted a chapter to The Principle of Similitude. He wrote, "Present day ideas of the resistance and propulsion of ships have been derived almost in to from the theories and methods evolved by the elder Froude, William Froude, dating back to about 1870. He first applied to such questions the Law of Comparison or Froude's Law, as it is well called, connecting the resistances of similar vessels."[51]

Taylor also wrote that the principle of similitude was first put forth by Newton, but Froude appeared to have developed the "particular form used to compare models of ships and of propellers, and to have been the first to use the Law of Comparison to obtain practical results."[52] Taylor further wrote that, "Froude's Law is a particular case of the general law of mechanical similitude, defining the necessary and sufficient conditions, that two systems or aggregation of particles which are initially geometrically similar should continue to be at corresponding times not only geometrically but mechanically similar."[53]

Taylor went on to cite work done on the "principle of dimensional homogeneity" by Dr. Buckingham of the Bureau of Standards and presented in his 1915 paper before the American Society of Mechanical Engineers. According to Taylor, "Froude's Law is readily deduced from this principle." And the "process developed by the elder Froude, now universally accepted, and used in model basin research the world over."[54]

Froude "played a leading part in interpreting the significance of Rankine's streamline ovals and formulated the streamline theory of resistance, and explained that the components of ship resistance were skin friction, wave-, and eddy-making and assigned approximate values of each."[55]

His 1869 "Explanations" provided a thorough and cogent summary of the state of hydrodynamic principles and vessel construction as they existed at the time.[56]

His work laid the groundwork for those who followed. To this day, Froude's work is a respected and essential fundamental of model testing for ocean, air, and spacecraft.

Froude's Theory on the Rolling of Ships, discussed in detail in this chapter, is another important contribution that stands to this day. Froude correctly identified the mechanics of a wave when he stated that the water remained stationary while the motion moved from one point to another. He concluded that excessive rolling of a ship was the result of a ship whose period corresponded to the period of the wave, which caused the cumulative effect to be greater than what would be expected from a single wave.

The Froude Number is an enduring result of the work Froude did at Torquay. This dimensionless number is calculated as the ratio of a body's inertia to gravitational forces and is still used when evaluating hull design. For an apt comparison of model to ship, both must be operating at the same Froude number. Calculations will be made to bring other factors into line when making design considerations. In general testing of models, the greater the Froude number, the greater the resistance. Engineers still rely upon this number today.

Notes

1. Brunel, I. (1870). *The Life of Isambard Kingdom Brunel, Civil Engineer*. London, Longmans, Gree, and Co.
2. (1955). *The Papers of William Froude M.A., LL.D., F.R.S. 1810–1879*. London, Institution of Naval Architects; Gawn, A. (1955). An Evaluation of the Work of William Froude. *The Papers of William Froude, M.A., LL.D., F.R.S. 1810–1879*. R. N. Duckworth, Captain A. D. London, The Instution of Naval Architects: xv-xxii.
3. Merrifield, C. W. (1870). The Experiments Recently Proposed on the Resistance of Ships. *Transactions of the Instutition of Naval Architects*. London, Institution of Naval Architects. *XI*: 80–93.
4. Barnaby, K. C. (1960). 1860–1864 The First Five Years - Iron Ships. *The Institution of Naval Architects, 1860–1960; an historical survey of the institutions transactions and activities over 100 years*. London,, Royal Institution of Naval Architects in association with Allen and Unwin: 7–47.
5. Merrifield, C. W. (1870). The Experiments Recently Proposed on the Resistance of Ships. *Transactions of the Instutition of Naval Architects*. London, Institution of Naval Architects. *XI*: 80–93.
6. Ibid.
7. Slade, S. (1998). "Towing Tank Tests." from http://www.navweaps.com/index_tech/tech-010. htm.

8. Gawn, A. (1955). An Evaluation of the Work of William Froude. *The Papers of William Froude, M.A., LL.D., F.R.S. 1810–1879*. R. N. Duckworth, Captain A. D. London, The Institution of Naval Architects.

9. (1955). *The Papers of William Froude M.A., LL.D., F.R.S. 1810–1879*. London, Institution of Naval Architects; Gawn, A. (1955). An Evaluation of the Work of William Froude. *The Papers of William Froude, M.A., LL.D., F.R.S. 1810–1879*. R. N. Duckworth, Captain A. D. London, The Instution of Naval Architects: xv-xxii.

10. Froude, W. (1861). On the Rolling of Shipsibid: 40–64; (1955). *The Papers of William Froude M.A., LL.D., F.R.S. 1810–1879*. London, Institution of Naval Architects.

11. Froude, W. (1861). On the Rolling of Ships. *The Papers of William Froude, M.A., LL.D., F.R.S. 1810–1879*. R. N. Duckworth, Captain A. D. London, The Instution of Naval Architects: 40–64; (1955). *The Papers of William Froude M.A., LL.D., F.R.S. 1810–1879*. London, Institution of Naval Architects.

12. Froude, W. (1861). On the Rolling of Ships. *The Papers of William Froude, M.A., LL.D., F.R.S. 1810–1879*. R. N. Duckworth, Captain A. D. London, The Instution of Naval Architects: 40–64; (1955). *The Papers of William Froude M.A., LL.D., F.R.S. 1810–1879*. London, Institution of Naval Architects.

13. Froude, W. (1861). On the Rolling of Ships. The Papers of William Froude, M.A., LL.D., F.R.S. 1810–1879. R. N. Duckworth, Captain A. D. London, The Instution of Naval Architects: 40–64; (1955). *The Papers of William Froude M.A., LL.D., F.R.S. 1810–1879*. London, Institution of Naval Architects.

14. Froude, W. (1861). On the Rolling of Ships. *The Papers of William Froude, M.A., LL.D., F.R.S. 1810–1879*. R. N. Duckworth, Captain A. D. London, The Instution of Naval Architects: 40–64.; (1955). *The Papers of William Froude M.A., LL.D., F.R.S. 1810–1879*. London, Institution of Naval Architects.

15. Froude, W. (1861). On the Rolling of Ships. *The Papers of William Froude, M.A., LL.D., F.R.S. 1810–1879*. R. N. Duckworth, Captain A. D. London, The Instution of Naval Architects: 40–64; (1955). *The Papers of William Froude M.A., LL.D., F.R.S. 1810–1879*. London, Institution of Naval Architects.

16. Froude, W. (1862). On the Rolling of Ships - Appendices. *The Papers of William Froude, M.A., LL.D., F.R.S. 1810–1879*. R. N. Duckworth, Captain A. D. London, The Instution of Naval Architects: 65–76; (1955). *The Papers of William Froude M.A., LL.D., F.R.S. 1810–1879*. London, Institution of Naval Architects.

17. Froude, W. (1862). On the Rolling of Ships - Appendices. *The Papers of William Froude, M.A., LL.D., F.R.S. 1810–1879*. R. N. Duckworth, Captain A. D. London, The Instution of Naval Architects: 65–76; (1955). *The Papers of William Froude M.A., LL.D., F.R.S. 1810–1879*. London, Institution of Naval Architects.

18. Froude, W. (1862). On the Rolling of Ships - Appendices. *The Papers of William Froude, M.A., LL.D., F.R.S. 1810–1879*. R. N. Duckworth, Captain A. D. London, The Instution of Naval Architects: 65–76; (1955). *The Papers of William Froude M.A., LL.D., F.R.S. 1810–1879*. London, Institution of Naval Architects.

19. Froude, W. (1868–1870). Observations and suggestions on the subject of determining by experiment the resistance of ships - Correspondence with the Admiralty. *The Papers of William Froude, M.A., LL.D., F.R.S. 1810–1879*. R. N. Duckworth, Captain A. D. London, The Instution of Naval Architects: 120–128; (1955). *The Papers of William Froude M.A., LL.D., F.R.S. 1810–1879*. London, Institution of Naval Architects.

20. Froude, W. (1868–1870). Observations and suggestions on the subject of determining by experiment the resistance of ships - Correspondence with the Admiralty. *The Papers of William Froude, M.A., LL.D., F.R.S. 1810–1879*. R. N. Duckworth, Captain A. D. London, The Instution of Naval Architects: 120–128; (1955). *The Papers of William Froude M.A., LL.D., F.R.S. 1810–1879*. London, Institution of Naval Architects.

21. Froude, W. (1868–1870). Observations and suggestions on the subject of determining by experiment the resistance of ships - Correspondence with the Admiralty. *The Papers of William Froude, M.A., LL.D., F.R.S. 1810–1879*. R. N. Duckworth, Captain A. D. London, The Instution

of Naval Architects: 120–128; (1955). *The Papers of William Froude M.A., LL.D., F.R.S. 1810–1879.* London, Institution of Naval Architects.

22. Froude, W. (1868–1870). Observations and suggestions on the subject of determining by experiment the resistance of ships - Correspondence with the Admiralty. *The Papers of William Froude, M.A., LL.D., F.R.S. 1810–1879.* R. N. Duckworth, Captain A. D. London, The Instution of Naval Architects: 120–128; (1955). *The Papers of William Froude M.A., LL.D., F.R.S. 1810–1879.* London, Institution of Naval Architects.

23. Froude, W. (1868–1870). Observations and suggestions on the subject of determining by experiment the resistance of ships - Correspondence with the Admiralty. *The Papers of William Froude, M.A., LL.D., F.R.S. 1810–1879.* R. N. Duckworth, Captain A. D. London, The Instution of Naval Architects: 120–128; (1955). *The Papers of William Froude M.A., LL.D., F.R.S. 1810–1879.* London, Institution of Naval Architects.

24. Froude, W. (1868–1870). Observations and suggestions on the subject of determining by experiment the resistance of ships - Correspondence with the Admiralty. *The Papers of William Froude, M.A., LL.D., F.R.S. 1810–1879.* R. N. Duckworth, Captain A. D. London, The Instution of Naval Architects: 120–128; (1955). *The Papers of William Froude M.A., LL.D., F.R.S. 1810–1879.* London, Institution of Naval Architects.

25. Froude, W. (1868–1870). Observations and suggestions on the subject of determining by experiment the resistance of ships - Correspondence with the Admiralty. *The Papers of William Froude, M.A., LL.D., F.R.S. 1810–1879.* R. N. Duckworth, Captain A. D. London, The Instution of Naval Architects: 120–128; (1955). *The Papers of William Froude M.A., LL.D., F.R.S. 1810–1879.* London, Institution of Naval Architects.

26. Froude, W. (1868–1870). Observations and suggestions on the subject of determining by experiment the resistance of ships - Correspondence with the Admiralty. *The Papers of William Froude, M.A., LL.D., F.R.S. 1810–1879.* R. N. Duckworth, Captain A. D. London, The Instution of Naval Architects: 120–128; (1955). *The Papers of William Froude M.A., LL.D., F.R.S. 1810–1879.* London, Institution of Naval Architects.

27. (1955). *The Papers of William Froude M.A., LL.D., F.R.S. 1810–1879.* London, Institution of Naval Architects; Gawn, A. (1955). An Evaluation of the Work of William Froude. *The Papers of William Froude, M.A., LL.D., F.R.S. 1810–1879.* R. N. Duckworth, Captain A. D. London, The Instution of Naval Architects: xv-xxii.

28. (1955). *The Papers of William Froude M.A., LL.D., F.R.S. 1810–1879.* London, Institution of Naval Architects; Gawn, A. (1955). An Evaluation of the Work of William Froude. *The Papers of William Froude, M.A., LL.D., F.R.S. 1810–1879.* R. N. Duckworth, Captain A. D. London, The Instution of Naval Architects: xv-xxii.

29. (1955). *The Papers of William Froude M.A., LL.D., F.R.S. 1810–1879.* London, Institution of Naval Architects; Gawn, A. (1955). An Evaluation of the Work of William Froude. *The Papers of William Froude, M.A., LL.D., F.R.S. 1810–1879.* R. N. Duckworth, Captain A. D. London, The Instution of Naval Architects: xv-xxii.

30. Froude, W. (1873). Description of a machine for shaping models used in experiments on forms of shipsibid.: 206–212; (1955). *The Papers of William Froude M.A., LL.D., F.R.S. 1810–1879.* London, Institution of Naval Architects.

31. Froude, W. (1873). Description of a machine for shaping models used in experiments on forms of ships. *The Papers of William Froude, M.A., LL.D., F.R.S. 1810–1879.* R. N. Duckworth, Captain A. D. London, The Instution of Naval Architects: 206–212; (1955). *The Papers of William Froude M.A., LL.D., F.R.S. 1810–1879.* London, Institution of Naval Architects.

32. Froude, W. (1872). Experiments on the surface-friction Experienced by a Plane Moving through Water. *The Papers of William Froude, M.A., LL.D., F.R.S. 1810–1879.* R. N. Duckworth, Captain A. D. London, The Instution of Naval Architects: 138–146.

33. Froude, W. (1877). Experiments upon the effect produced on the wave-makiing resistance of ships by length of parallel middle body. *The Papers of William Froude, M.A., LL.D., F.R.S. 1810–1879.* R. N. Duckworth, Captain A. D. London, The Instution of Naval Architects: 311–319.

34. Ibid.

35. Ibid; (1955). *The Papers of William Froude M.A., LL.D., F.R.S. 1810–1879*. London, Institution of Naval Architects.
36. Froude, W. (1877). Experiments upon the effect produced on the wave-makiing resistance of ships by length of parallel middle body. *The Papers of William Froude, M.A., LL.D., F.R.S. 1810–1879*. R. N. Duckworth, Captain A. D. London, The Instution of Naval Architects: 311–319.
37. Ibid.
38. Ibid.
39. Ibid.
40. Ibid.
41. Ibid.
42. Ibid.
43. Ibid.
44. Ibid.
45. Ibid.
46. Froude, W. (1874). On Experiments with H.M.S. Greyhound. *The Papers of William Froude, M.A., LL.D., F.R.S. 1810–1879*. R. N. Duckworth, Captain A. D. London, The Instution of Naval Architects: 232–253.
47. Ibid.
48. Ibid.
49. Ibid.
50. Ibid.
51. Taylor, D. W. (1943). The Principle of Similitude. *The speed and power of ships; a manual of marine propulsion*. Washington,, U. S. Govt. Print. Off.: ix, 301 p.
52. Ibid.
53. Ibid.
54. Ibid.
55. Gawn, A. (1955). An Evaluation of the Work of William Froude. *The Papers of William Froude, M.A., LL.D., F.R.S. 1810–1879*. R. N. Duckworth, Captain A. D. London, The Institution of Naval Architects.
56. Froude, W. (1869). The State of Existing Knowledge on the Stability, Propulsion and Sea-going Qualities of Ships, and as to the Application Which it may be Desirable to Make to her Majesty's Government on this Subjectibid. The Instution of Naval Architects: 129–133.

Chapter 7
David Watson Taylor

I never knew anyone who I thought had such a fine mind, was so broad as well as practical in all matters, had such clear foresight and, withal, was so gentle and lovable at all times. He had, also, that truly rare gift of being able to answer a question without making the inquirer feel that he should have known the answer.

Rear Admiral George H. Rock on Rear Admiral David Watson Taylor

As the start of the nineteenth century drew near, the United Kingdom had a longstanding interest in a strong navy and an investment in that navy that included the work of William Froude. An island nation with colonies and possessions around the globe, she recognized the importance of being able to reach and strike outside her boarders swiftly and definitively. Since many European countries also had colonies and possession to consider, a strong navy was equally important to them. The United States, previously concerned with intercoastal waterways and the defense of her shores, had a newly born and burgeoning interest in both being perceived as a reliable ally and having the ability to trade easily with overseas partners. She was also coming to the realization that to be a world-player, she would need a world-class navy.

Unfortunately, as the Civil War ended in 1865, most of America's ships were laid up in reserve. The condition of the Navy continued to deteriorate and by 1878 it had only 6,000 men. By the 1880s there was a push for a "New Navy." This New Navy would be a blue-ocean navy with vessels that could cross the ocean to come to the aid of allies and bring the merchants of the United States access to the markets of Europe and Great Britain.

The first ships of the New Navy were authorized in 1883. Known as the A-B-C-D ships they were three Protected Cruisers (so-called because of the armored deck to protect essential machinery beneath the deck), and one smaller "dispatch" boat (to carry messages between the larger ships). The next year, the Naval Advisory Board recommended the construction of a model basin, so the United States could

G. Hagler, *Modeling Ships and Space Craft: The Science and Art of Mastering the Oceans and Sky*, DOI 10.1007/978-1-4614-4596-8_7,
© Springer Science+Business Media, LLC 2013

test the design of new ships before construction was begun. Because these new ships would be costly and were vital to the interest of the United States, it was felt that anything proven to cut costs in the long run would be of benefit. Even with the history of successful work at model basins in England and other countries, it would be 12 more years before work on the basin in the United States would actually begin.

During the period 1885–1890, "Congress and the Navy sorted through a variety of budgetary and technical questions, none of which had simple answers. Members of Congress of both parties hoped to achieve improvements at the lowest possible cost, and evaluated each technical change against its impact upon federal expenditure. To what extent should armor be sacrificed for speed, especially in cruisers? How many engines and screws should be used to propel those most powerful, heaviest ships? Should double- or triple-expansion engines be used? How should propellers be modified to improve their efficiency?"[1]

The publication of *The Influence of Sea Power Upon History, 1660–1783*, by Alfred Thayer Mahan in 1890, as mentioned earlier, proved to be the tipping point for a global interest in strong navies that could be part of a "decisive battle." Mahan's case was so convincing that it has influenced international naval strategy into the current century. He wrote that to be effective, a nation must have dependable, battle-worthy ships; at that time, armored ships powered by steam. The need for more powerful steam-powered metal ships had never felt as urgent. The ships must be built quickly, economically, and successfully.

Congress and the Navy wanted to use a more scientific approach to design these new ships. They wanted a process that would make it possible to know as early as the design phase, if a ship would perform as expected, and what configuration of power and structure would result in the best performance. Through the efforts and advocacy of American Naval Architect David Watson Taylor, the United States was aware of the advances in ship design that had been brought about in England by the use of Froude's model testing. They understood they could run model tests to perfect designs for ships of the fleet prior to construction. They also understood they needed to build a testing facility before this work could begin. In 1896, it was decided that the United States would have a model basin. Taylor was chosen to design, construct, and head the Experimental Model Basin.

David Watson Taylor

A brilliant mathematician, Taylor attended Randolph-Macon College at age 13. The youngest boy in the college, he was appointed to the United States Naval Academy, then the premier institute for mathematics, upon graduation in 1881. In 1885 he graduated as a cadet engineer, with the highest academic record to date. While at the Naval Academy, he was active in a number of activities and in his fourth year, he was a member of the football and baseball teams, president of the athletic association and chairman of the "hop" committee. Upon graduation he served 3 months on the U.S.S. *Pensacola*, the flagship of the European Squadron, commanded by Captain George Dewey[2] before the US Navy assigned him to study the design and construction of warships and naval machinery at the Royal Naval College in Greenwich, England. Study in England or Europe was necessary because the United States then had no program.

Taylor entered the postgraduate course in Greenwich and specialized in marine engineering. In 1888 he graduated with a first-class certificate, and again achieved the highest grades to that time. In 1885, while at Greenwich, he was appointed Assistant Naval Constructor due to his high standing at the College.

When he returned to the United States, Taylor was assigned to duty at Cramp's shipyard in Philadelphia. In 1889 he was a member of the board of experts considering alleged defects in the battleship *Texas* being built at Norfolk. At about that time he also assisted in preparing the designs of naval vessels for the consideration of the *Board of Naval Policy*. Taylor also took an active part in the 1880s "New Navy," program during which a number of vessels were constructed for "coast defense."[3]

Taylor was promoted to the grade of Naval Constructor in 1891, and from 1892 to 1894 he served as Construction Officer in the Navy Yard, Mare Island, California. In 1893, Taylor's first book, "Resistance of Ships and Screw Propulsion," was published. In 1894 he was assigned to duty in the Bureau of Construction and Repair at Washington as principal assistant to the Chief Constructor, Philip Hichborn. Early in his career, Taylor "critically investigated the various methods of 'ship calculation' for the determination of displacement as well as the characteristics of buoyancy and stability and formulated a method of calculation which became the standard for the Navy."[4]

It was with this impressive body of experience and a deep respect for Froude's work, that Taylor was ready at age 32, to turn his scrupulous attention to the task of building the ultimate testing facility at what is now the Old Navy Yard in Washington, DC. It would take 3 years to complete the facility. Despite the complexity of the task and several unanticipated problems, he would bring the project in on budget.

The Experimental Model Basin

At the eighth general meeting of the Society of Naval Architects and Marine Engineers, held in New York, November 15 and 16, 1900, Taylor, then Naval Constructor, presented his report, "The United States Experimental Model Basin." This report gave a complete picture of the obstacles and solutions along the way to completion of the Basin, as well as the results of the testing done since its opening in 1898.

The Experimental Model Basin (EMB) was located in the southeast corner of what is now known as the Old Navy Yard in Washington, DC. It consisted of a basin and the building that housed it (Figs. 7.1 and 7.2). The models were moved through the water with a towing carriage much like the one used by William Froude (Fig. 7.3).

"Circumstances necessitated the location of the basin upon a site which left much to be desired, requiring a much more expensive method of construction than would have been necessary had conditions been more favorable," Taylor wrote in his report. "As will be seen from the plans, the basin is very close to the Eastern Branch

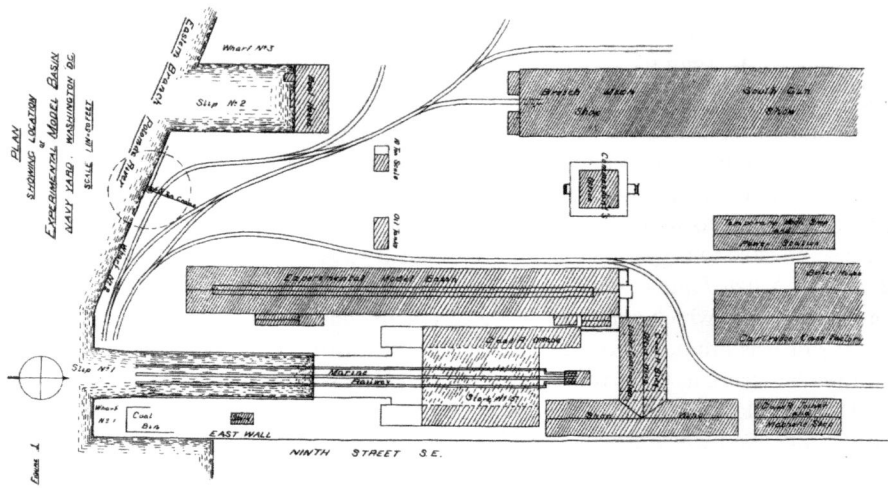

Fig. 7.1 EMB plot plan. The experimental model basin and surrounding land. *Source*: "The United States Experimental Model Basin" by David W. Taylor, 1900

Fig. 7.2 EMB photo. The experimental model basin at the Old Navy Yard. *Source*: National Archives

Fig. 7.3 The towing carriage. *Source*: "The United States Experimental Model Basin" by David W. Taylor, 1900

of the Potomac, and has under it several weak springs. Indeed, during construction there was much apprehension of quicksand toward the northern end."

To ensure the basin would not collapse from the water pressure building beneath it, Taylor built it with sheet piling around it. Nearly 3 years after all concrete work had been completed, Taylor wrote that, "there has been no trouble from leakage, and no indication of settling or unsoundness has developed at any point, it is believed that the difficulties of foundation may be considered to have been successfully overcome." Still, whenever the basin was pumped dry, a job that took about 4 h, "the outside water is pumped once a day in order to keep it from rising sufficiently high to bring serious upward pressure on the bottom."[5]

With the basin itself constructed and stable, the next issue was filling it with one million gallons of clear, debris-free water, and keeping the level constant. The water itself was taken from the Potomac River. On the way to the Basin it was "treated with a minute quantity of alum, which coagulates with any mud present, and then clarified by passage through a sand filter of the pressure type." At full capacity, it took about a week to fill the basin; more if the Potomac River was unusually muddy. Once the basin was filled, Taylor wrote a "small stream is kept constantly running through the filters in order to freshen the water, make up for leakage, waste, etc."[6]

The building was heated and cooled by use of an ingenious system. In winter, hot air was drawn through steam coils by an electric fan. "Thermometers are fitted in connection with a small compressed air installation," Taylor wrote, "by which registers are automatically closed and windows opened, or vice versa, as necessary to

Fig. 7.4 Wave breakers at the end of the towing tank. *Source*: "The United States Experimental Model Basin" by David W. Taylor, 1900

maintain the temperature evenly at a desired point." Because the air was damp and the people working in the basin were involved in moving rapidly on the traveling carriage that towed the models, the temperature was kept "slightly higher than that desirable for an ordinary living or working room."[7]

Although the models attached to a towing carriage that sat above the water and ran along tracks on the rims of the basin before they were moved through the water, disruptions in the water in the form of waves generated by each test run were still a potential problem. To dissipate the waves as quickly as possible, Taylor designed steel troughs on each side that were just below the surface of the water and absorbed the wave disturbances. At the northern end of the basin, he had a wave breaker "consisting of a large number of square strips of wood set vertically at varying distances apart." The combination of the steel troughs and the wooden wave breakers was enough to "give entire satisfaction," Taylor wrote. "Without wave-breaking appliances it would be a very lengthy operation to run trials at high speeds as very long waits between runs would be necessary in order to allow subsidence of waves."[8] (Fig. 7.4).

Taylor mounted the equipment for determining the resistance of the model on the traveling carriage that spanned the basin. The carriage weighed about 70,000 pounds. As a result, it could be relied upon to maintain a constant speed once it reached the desired speed. The carriage was driven by a total of four motors, one on each corner. The speed was controlled by the Ward-Leonard system that used an "exciter" generator to keep the field coils of the motors in constant excitation. "Current from the same exciter is passed through the controlling rheostats on the

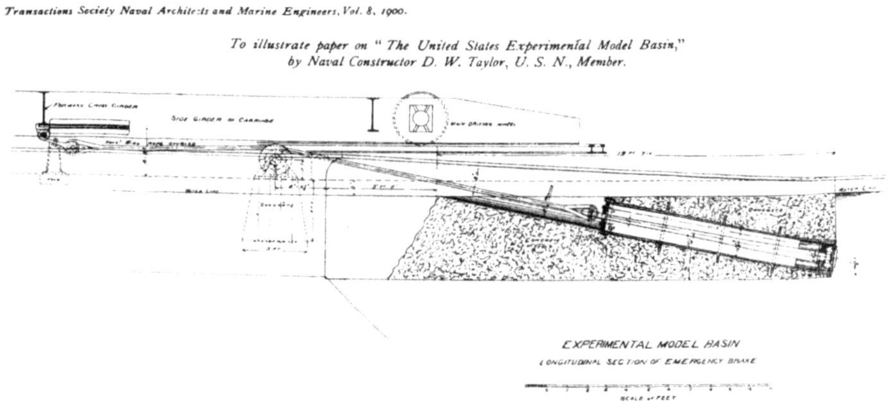

Transactions Society Naval Architects and Marine Engineers, Vol. 8, 1900.

To illustrate paper on " The United States Experimental Model Basin,"
by Naval Constructor D. W. Taylor, U. S. N., Member.

EXPERIMENTAL MODEL BASIN
LONGITUDINAL SECTION OF EMERGENCY BRAKE

SCALE IN FEET

Fig. 7.5 The emergency breaking system. *Source*: "The United States Experimental Model Basin"
by David W. Taylor, 1900

carriage and also around the field coils of the main generator," wrote Taylor. With
the main generator "kept running at a constant speed by a governor which limits
variation of speed within $1^{1/2}$ percent from no load to full load… the result in practice is to give the carriage excellent speed control and regulation."[9]

Taylor had his basin filled with clear, fresh water, an apparatus for measuring
resistance and a towing carriage that would move steadily at the required speed.
"With such a heavy mass moving at this speed in a confined space," he wrote, "it was
necessary to devise the arrangements for stopping with the utmost care." With his
characteristic attention to detail, Taylor devised three separate braking systems.[10]

The first took advantage of the Ward-Leonard system. "With the Ward-Leonard
system," he wrote, "a very powerful electrical braking effect is obtained from the
driving motors, through the back current which they generate when the exciter current around the generator fields is shut off or reversed." This method allowed the
towing carriage to be stopped fairly quickly but it had some serious drawbacks.
"Clearly," Taylor wrote, "however, this method of stopping could not be relied upon
alone, since it fails if the circuit is broken either accidentally or by the automatic
circuit breakers in case of an overload." It also required the person on the towing
carriage to take action. "It is evidently desirable, then," Taylor wrote, "to have at
least one method of braking which shall stop the carriage in the minimum possible
distance, be independent of the person operating the electric current, and require no
manipulation on the part of the person operating the carriage."[11]

A friction brake was decided upon as the second system. It would be closed by
hydraulic pressure and the pressure could be read by a gauge at the south end of the
basin so that, "operators could be certain before starting a run that the friction brake
has pressure on." Also, "The apparatus was designed for a maximum pressure of
600 pounds to the square inch," Taylor reported, "It is found, however, that a pressure of 300 pounds is sufficient to bring the carriage to rest from its maximum speed
within less than 20 feet."[12]

"As a final provision for safety," Taylor wrote, "there is fitted what is called the emergency brake, which takes hold of the carriage if in any way it gets through the friction brake without being arrested." With this braking system, "the carriage engages a large hook connected by heavy cables and a taper piston rod to a piston working in a hydraulic cylinder." It had never been necessary to use this emergency brake at the time of Taylor's report. "It is hoped that this emergency brake will never be called upon to demonstrate its capacity," he wrote.[13] (Fig. 7.5).

The Models

The assumption was that the new EMB would use models of the same type and size as those used by Froude. Unfortunately, this proved to be impractical. Taylor wrote in his report, "Parrafine presents many advantages for models, but for us has the unsurmountable disadvantage that it will not stand the summer temperature in Washington without inadmissible softening." Wood models of white pine were chosen because they retained shape during changes in weather and were overall stronger than paraffin. Taylor's report mentioned three disadvantages that had been overcome. The first two were that wood models were difficult and expensive to shape, and harder to keep tight. These were "practically overcome by the adoption of special machinery." The last disadvantage was that it was harder to achieve a uniform surface. This was overcome "by using a special varnish to finish the models, which gives a surface practically uniform," wrote Taylor.[14]

Froude's paraffin models had been 12 or 14 feet in length. "Owing to the greater strength of wood it appeared perfectly feasible to make models 20 feet long," Taylor wrote, "and the sectional area of the basin was such that these models could be run with no greater interference owing to limited size of basin than 12-foot models in the smaller foreign basins."[15]

Taylor soon discovered there were advantages to the 20-foot models when it came to determining resistance. He wrote, "We find that for the 20-foot models of practically all our naval vessels the resistances at the speeds corresponding to the actual maximum speeds of the vessels are below 40 pounds. With 12-foot models the resistances would have been below 9 pounds." He went on to report, "With the large model resistance is accurately measured to a given percentage of accuracy with less difficulty, and the gap between model and ship to be bridged by the law of comparison is not so great."[16]

Confident that the 20-foot white pine models would provide more accurate results, Taylor went on to describe the five resistance curves that were obtained for each model: "No. I. With the model at a displacement corresponding to the designed normal displacement of the ship and at the designed trim of the ship. No. 2. With the model as in No. I, except the trim is changed 4 inches by the head. No. 3. With the model as in No. I, except the trim is changed 4 inches by the stern. No. 4. With the model as in No. I, except that it is 10 percent lighter. No. 5. With the model as in No. I, except that it is 10 percent heavier." Running tests on models with the same five modifications gave Taylor a solid basis for comparing the performance of different designs.[17]

Making the Models

The first step in making a 20-foot model was to take the body plan drawing and enlarge it to determine the "sections of a 20-foot model corresponding to the sections in the body plan," Taylor wrote in his report. An enlarging pantograph or ediograph was used for this purpose (Fig. 7.6). Once the sections were enlarged, they were used as patterns for the construction of a "former model" made of "a skin of strips of wood nailed securely to them and smoothed off."[18]

While the former model was being built, "a wooden block is built up of white pine lifts about two inches thick, glued together hot under heavy hydraulic pressure," Taylor wrote. "This block is so proportioned that when the finished model is cut from it the wood will be left amply thick, nowhere less than about two inches. Additional thickness is not avoided since the models require ballast in every case."[19]

Once model makers had the former model and a white pine block of the required thickness, the next step was to cut the actual 20-foot model. The former model and the corresponding block of white pine were secured in the model-cutting machine with the former model being below. "The roller below rolls over the former model, and the saw above, which is driven at about 2,200 revolutions per minute by an electric motor," wrote Taylor, "is constrained by the balanced link work to move exactly above and at a uniform distance from it. The sizes are so arranged that the saw does not cut within one-eighth of an inch of the intended finished surface of the model."[20] (Figs. 7.7, 7.8, and 7.9).

Fig. 7.6 The ediograph. *Source*: "The United States Experimental Model Basin" by David W. Taylor, 1900

Fig. 7.7 The cutting machine 01. *Source*: "The United States Experimental Model Basin" by David W. Taylor, 1900

A rotary cutter was then attached to the machine to rough finish the model. The final sanding was done by hand with sanded disks driven by an electric motor. The models were then painted inside and out. The final step was the application of a standard varnish on the outside.

When needed for a test, the model would be suspended from one of the cranes for that purpose and weighed. Ballast would be added until the weight in the water corresponded to the desired displacement of the ship it represented. When the test was completed, the model would be weighed again when the ballast was removed. Then the model would be stowed on the galleries on each side of the basin.[21]

Results Reported in the 1900 Report

Taylor was able to report that the curves so far tended "to strengthen the theory that length and displacement are the primary factors involved in resistance; or, in other words, that given the length and displacement of vessels of usual forms, the resistance is not materially changed by practicable changes in shape, beam, draught, etc."[22]

Some early curves showed what Taylor referred to as "peculiar 'humps'" at between four and five knots. They were most pronounced in narrow, deep models and died away and appeared as flat spots for the broad, shallow models. Taylor reported that, "they are, of course due to the interference between the bow and stern wave systems."[23]

Fig. 7.8 The cutting machine 02. *Source*: "The United States Experimental Model Basin" by David W. Taylor, 1900

Taylor's report included a discussion of curves for several battleships before concluding, "Some results obtained in connection with the Georgia models show conclusively the direct practical value of model experiments." He made this observation because through model testing, he had been able to compare differences resulting from different shapes in two models in the Georgia class. Taylor wrote of the results, "Although we knew in a general way that speed would be gained with increased length in spite of greater displacement, there is no reliable method known at present by which we could possibly have determined satisfactorily the necessary and sufficient length without model experiments."[24]

Fig. 7.9 The cutting machine 03. *Source*: "The United States Experimental Model Basin" by David W. Taylor, 1900

Because US battleships required a shallow draught and fuller forms, they were more difficult to drive than a foreign battleship with a deeper draught and the same displacement. "Speed can be obtained by increased length," Taylor wrote, "but, as the members of this Society are well aware, great length is peculiarly objectionable in a battleship…" "It was evidently, then, a matter of the greatest practical importance to be able to determine satisfactorily," Taylor wrote, "such a vital feature as the necessary and sufficient length, and the requisite horse-power for our five new battleships, which will represent, completed, and investment of some thirty millions of dollars." Thus, the tests done in the early years of the model basin were instrumental in arriving at the proper length and horsepower for the new battleships.[25]

Work with Elmer Sperry

Taylor was interested in anything that could increase the efficiency of the fleet. From the type of glue used for his models to instruments that could be used aboard the ships themselves, Taylor wanted to examine the idea and test it thoroughly.

When Elmer Sperry came to him in the winter of 1907 with a device that might be used to reduce the roll of ships at sea, Taylor launched in to a period of experimentation and consultation with Sperry. The idea of Sperry's gyroscopic stabilizer was that it would begin to stabilize the roll of the ship before the full impact of the wave was even felt. "Taylor agreed to test a model of Sperry's stabilizer at the basin, and in July 1910 he drafted a forty-page report explaining in mathematical terms the stabilizing effect of Sperry's machinery."[26] He then suggested to Sperry that a gyrocompass might also be of use.[27]

The Suction of Vessels

In 1909, Taylor published the results of his study of suction between two large ships passing in a narrow channel in a paper, "Some Model Experiment on Suction of Vessels." The paper, based on the research performed in the EMB, was presented to the Society of Naval Architects and Marine Engineers in New York. Taylor had been interested in "… the question of 'suction' or reduced pressure that occurs between two ships, traveling in the same direction, when one overtakes and passes the other too closely. He saw the problem as a demonstration of the principle of streamline flow, in which he had been interested since the 1890s. As early as 1902 Taylor explained the principles based on streamline theory over and over in his correspondence."[28] His paper demonstrated "… how reduced pressure between closely passing ships tended to force them closer together with danger of collision."[29]

Taylor had given thought to the question of suction for some time before the publication of his paper. On November 29, 1902, he wrote to Mr. W.H. Faust at the US Naval Recruiting Station, "Replying to yours of the 25th. There is no question that the kind of suction to which you refer exists in the case of ships meeting or passing one another quite closely. It would be the natural supposition that taking two ships side by side and quite close together in the water where they approach each other nearest, that is to say where there is less space between them, there would be a heavier pressure than elsewhere, which would tend to force them further apart. As a matter of fact the stream line theory of motion of fluids, which is amply confirmed by experiments, shows that the smaller the opening the less the pressure, and hence there is a suction. Towards the bows of the ships there would be a pressure tending to force the bows apart, and this combined with the suction amidships, is quite able to give a very rank sheer in many cases. I have known of one or two cases where suction was blamed unjustly, but in the case you describe I should think there would be no question that suction was the dominant factor in causing the collision. If I have not made myself clear please let me know and I will be glad to supplement this."[30]

As a result of his work with this theory, Taylor was called as an expert when the British cruiser *Hawke* collided with the White Stare liner *Olympic*. The Admiralty called on Taylor's expertise to support their claim that the collision was not the result of incompetence on the part of their captain.

The Titanic Investigation

Taylor was also part of the panel investigating the Titanic disaster. He was described at the time as, "Naval Constructor David Watson Taylor, U.S.N., is regarded as one of the foremost authorities on ship construction in the world. He has the unusual distinction of having been graduated by two of the greatest naval schools—the U.S. Naval Academy and the Royal College at Greenwich, England—after having made the highest marks in his examination that had ever been attained by a student in the history of either institution."[31]

In his remarks on the Titanic, Taylor wrote, "The *'Titanic'* catastrophe teaches no new lesson as regards the fallibility of man. It simply furnished another example of the well established principle that if in the conduct of any enterprise, an error of human judgment or faulty working of the human senses involves disaster, sooner or later the disaster comes.

Looking backward it is easy to see that the long established passage lanes of the Atlantic involved danger of just such an accident, and from the point of view of safety it was an error of judgment to give them such a northerly location.

Looking backward it seems an error of judgment of the captain of the *'Titanic'* to risk passage near the ice. The gallant officer and gentleman went down with his ship to honorable death, and his story can never be told. It seems practically certain that he did not for one moment think he was running any material risk of accident to his vessel, much less risk of destruction. The mere fact that he was not on the bridge at the time of the collision is very strong evidence he thought his course would have cleared the bergs whose position had been reported to him.

Picked captains of Atlantic liners cling to the bridge to the point of exhaustion whenever they consider circumstances to involve the least danger to the ship.

If Captain Smith erred, it was the error of a captain whose record and experience were of the best. We need not expect to secure grater safety by better captains, and without speculating as regard matters involving personnel and discipline, let us now consider matters of material.

The most salient fact is that if the *'Titanic'* had carried more boats or a number of life rafts in addition to her boats, many more lives would have been saved.... The facts that under the circumstances more boats would have saved many more lives from the *'Titanic,'* and that she should have carried about three times as many boats as she had should not blind our eyes to the fact that lifeboats are, after all, a very inefficient device for saving life from a sinking vessel...

Lifeboats, no matter how much improved, will probably always be inefficient as life-saving appliances for the mammoth steamers of today. Something different is needed..."[32]

The Speed and Power of Ships

The Speed and Power of Ships, published in 1910, was the seminal book in the field of vessel design. It brought together the current state of the science, including the work done by Froude. Organized into sections called books, it began with general information that included stream lines and a thorough discussion of the Principle of Similitude. The second book discussed all forms of resistance with reference to the work done by Froude and Taylor. Book III tackled propulsion, including the issues associated with cavitation, with a brief mention of jet propulsion as well. Book IV gave a thorough explanation of trials and their analysis. Included in this section is a notably nontechnical and expectedly detailed explanation of just what to look for in a measured mile course. This was an important information because many tests were run in the manner and the ability to compare results was a prime consideration. The final book was devoted to the powering of ships and the use of the Standard Series.

It was a complete treatment of the topic and typically, Taylor took a no-nonsense approach to its publication. In a letter to John Wiley & Son publishers dated March 2, 1910, Taylor wrote "Gentlemen: As per instructions of Major W.H. Wiley, I am forwarding you today by Adams Express, prepaid, manuscript and blue prints of figure of a work on resistance and propulsion of ships which I am proposing to call 'The Speed and Power of Ships.'

The text will make 275–300 octave pages as in Peabody's Naval Architecture. There are 12 sheets of table to Table XV and 108 sheets of figures to Fig. 276. I have tried in this work to cover completely the present state of science as regards resistance and propulsion going fully into the theory that the book may be suited to the use of students but given particular attention to methods, rules and formulae which will be of use to practicing naval architects and marine engineers...

Owing to the number of people in Great Britain interested in shipbuilding and marine engineering I believe that there would be a much larger sale for this work in England than in the United States... Vey respectfully, D.W. Taylor, Naval Constructor, U.S.N."[33]

The book laid out the work behind and the results of what is now referred to as The Taylor Standard Series. The series came about as a result of Taylor's desire to develop "...a practical shortcut to finding the best hull form—that is, the development of a series of like forms, each slightly varied from the next. ... He sought to isolate the factors affecting resistance and to develop curves of predicted effects known from experiments."[34]

"By 1908 Taylor worked from 'parent forms' in which certain factors would be held constant and others varied. ... Using parent forms, he tested three series of models with each series containing twenty models. Each series consisted of four different-sized models, each size with five different models with different curves of sectional area. Each of the three series of twenty models held to a single longitudinal coefficient. The experiments yielded optimum percentages of parallel middle body for each of the three longitudinal coefficients."[35] (Fig. 7.10).

Taylor's work on the optimal bow shape was part of his series work. He had published some finding in an earlier article, "Influence on the Bulbous Bow on Resistance," where he wrote "we were never able to make a swan model that drove more easily than the usual form but the bulbous bow rather grew out of this research. Its idea was to fine the water line as much as possible and put the displacement removed from the vicinity of the water line as far below water as possible.

So instead of a fine bow forward with a second full bow abaft it, we had a fine bow at the water line as it were and a full bow near the keel."[36] The bulbous bow is now used on every major vessel in the world.

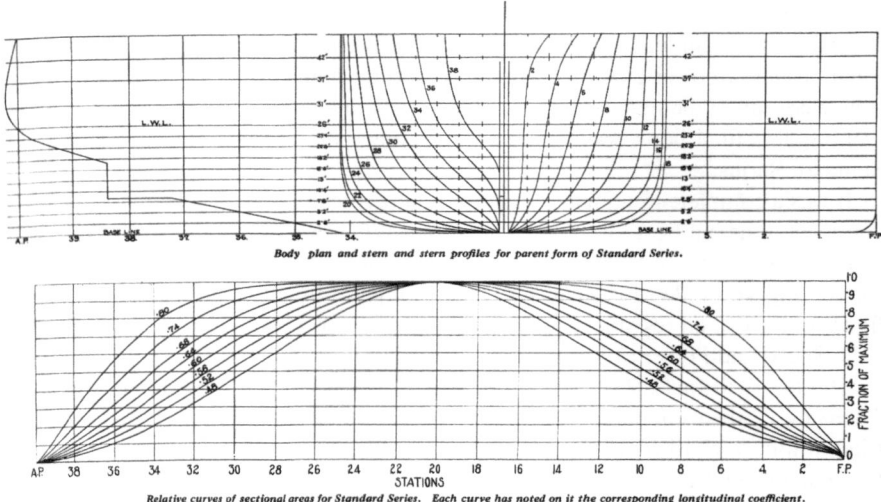

Body plan and stem and stern profiles for parent form of Standard Series.

Relative curves of sectional areas for Standard Series. Each curve has noted on it the corresponding longitudinal coefficient.

Fig. 7.10 The parent forms. *Source:* "The United States Experimental Model Basin" by David W. Taylor, 1900

Fig. 7.11 Bulbous bow. The bulbous bow extends forward of the hull, beneath the water line

"The conclusion that 'bulbous bows' rather than sharp bows can be better suited to high-speed ships emerged as the most widely recognized and revolutionary conclusion of Taylor's series work."[37] By having a bulbous, rounded bow, beneath the surface (Fig. 7.11), the water would go up and over the bulbous bow, with far less of the water rising up the sides of the bow to impede forward motion (Fig. 7.12).

Until that time, it was assumed that a sharp bow would meet the least resistance. Taylor proved that a rounded bow was more efficient in reducing resistance. As a result of his work, Taylor concluded that a bulbous bow below the waterline

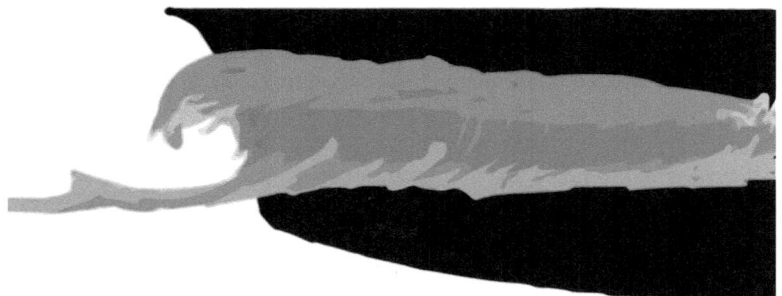

Fig. 7.12 Bow wave. The bow wave is unimpeded without a bulbous bow

would result in the best performance. He wrote, "At the high speeds... the whole forebody has to do with the creation of the waves, and we decrease resistance by making the waterline as fine as possible and putting as much as possible of the displacement well below water where the pressure due to its thrusting itself into undisturbed water will be a much as possible absorbed in doing the necessary pumping aft of the water and not in raising the surface into waves."[38]

Taylor's conclusion was based in part on his work with tracing streamlines. To observe streamlines, Taylor developed a method for making them visible. "The model hull would be coated with an iron compound set in glue. During model runs, small amounts of diluted pyrogallic acid would be injected through minute holes from inside the model, to flow out into the passing water. The acid would stain the hull, leaving a streak at a distance of two to four feet from the hole. After a short run, the model would be removed, and a new hole drilled to continue the streamline. The laborious process required half a day of testing to determine six lines of flow on a standard twenty-foot model. This simple demonstration, however, allowed Taylor to come to a major conclusion. Water was not 'parted' at the bow to flow out to the sides, as most mariners assumed, but rather flowed down and under the vessel. When combined with later observations, this point would allow Taylor to suggest major modifications in the bows of high-speed ships."[39]

The Speed and Power of Ships was updated in 1933, when 40 newer models were added to the series. The Taylor Series formed the basis for initial estimates on the speed of a ship given the resistance inherent in the design. Now incorporated into the computer models that are used for preliminary design activities, it informs the construction of ships to this day.

Aeronautics

By 1910, Taylor saw aeronautic research as a natural outgrowth of the work done at the EMB. He wanted the testing of these machines under the auspices of the EMB rather than the Smithsonian administration or the Bureau of Standards. The question of who would head aeronautic research was hotly contest from 1911 until 1915, when Congress created a formal advisory committee. "The National Advisory Committee for Aeronautics (NACA), modeled on a British advisory committee, in turn created an array of subcommittees to issue reports on specific aeronautic research topics and to arrange a few preliminary research contracts. Taylor was an advocate of NACA and was appointed to the committee in 1917 as a replacement for charter member Holden C. Richardson."[40]

Taylor and his assistant, William McEntee, were busy during this time. "In early March 1912 Taylor began planning construction of the wind tunnel at the navy yard. The proposal to construct the wind tunnel received quiet approval in the Bureau of Construction and Repair: neither Taylor not Bureau Chief Watt sought special congressional endorsement or funding, which might have proven difficult in the face of the ongoing national debate over the proper location for aeronautic work.

Through 1913, Taylor proceeded with construction of the foundation and the tunnel walls, and worked on details of the measuring devices, securing the aid of the Office of Naval Intelligence to inquire into purchase of plans for a British torsion balance. By July 1914 Taylor had completed calibration and regulation work, and began to look for a 'man to keep on aeronautic work only.' He wrote to Jerome Hunsaker, a young officer interested in aeronautics while he was still in his naval architecture course at MIT. Hunsaker was later to work for Taylor in the field of aeronautics, and become famous as a pioneer in naval aeronautics. In later life, Hunsaker would chair the National Advisory Committee for Aeronautics from 1941 to 1956...

Like the model basin, when the tunnel as completed it was the largest in the world, and it incorporated several departures from earlier precedents."[41] For one thing, the tunnel was designed so the exhaust air would return to the intake. For another, "the air speed was checked with an array of twelve Pitot tubes, which had been calibrated both at MIT and at Britain's National Physical Laboratory."[42]

"One of the better-known contributions of the aeronautics group at the Bureau of Construction and Repair during the period was the design and testing of the 'NC' aircraft. Recommended by Taylor during the war, these 'flying boats' or seaplanes were designed by Richardson and Hunsaker, along with others, to be flown across the Atlantic for deliver in Britain in case submarine attacks by the Germans interdicted surface transport. Although the war ended before the planes were completed, three were manufactured and flown, one by Captain Richardson. While he and another pilot had to bring their craft down in seas too rough for takeoff, a third, the NC-4 made it to the Azores and Great Britain in the historic first flights across the Atlantic, May 1919, eight years before Charles Linbergh's solo, nonstop flight in the *Spirit of St. Louis*."[43]

Through the efforts of Taylor and McEntee, aerodynamic research remained part of the work of the model basin.

NACA

Given Taylor's intellect and curiosity, it was only natural that he would take an interest in the new possibilities brought about by the Wright brother's success at Kitty Hawk in 1903. When development of the NC-type aircraft began in 1918, Taylor aided in the development. In 1919, the NC-4 was the first aircraft to make a transatlantic flight. Curtiss Aeroplane and Motor Company built these flying boats. They utilized a biplane design and were built with the intent of flying across the ocean so as to avoid the German subs that patrolled there. These flying boats took off from and landed on the water.

Once retired, Taylor served on many National Advisory Committee for Aeronautics (NACA) committees after its founding in 1915. His work with NACA gave him the opportunity to apply his expertise with hydrodynamic principles to the challenges of aerodynamics. In 1927 he became chairman on the Subcommittee on Aerodynamics. His special areas of interest were with the design of propellers, seaplane floats, and flying-boat hulls, all directly related to the work he had done with naval vessel propeller and hull design.

In this capacity he wrote to Orville Wright on February of 1925. "I am studying some airplane propeller questions, and am inclined to think that if the propellers on your original airplane had not been more efficient than many of those used to-day you would never have been able to fly. Would it be possible for you to let me have any plans of your original propellers? I would like to find out the diameter, pitch ratio, blade area, power, revolutions in normal flight, and nature of the blade sections.

I am afraid you could not give me all of this, but hope you will be able to give me some, as I wish to point the moral that the wonderful progress of aeronautics has not all been in the direction of efficient propellers."[44]

Orville replied on March 14, 1925. "I have delayed answering your letter until I could make a search through our records for the information you desired on our early propellers. I have failed to find the drawings of the 1903 propellers. I still have the propellers themselves, but they are stored where I cannot get at them at present. I feel sure that I have all of this data as it was written at the time, but I am sorry to say that these old papers are not now in order for easy reference. I will continue the search and hope to get this information for you later.

In 1903 we calculated the efficiency of our propellers at about 66 percent. The flight of the machine seemed to corroborate these calculations or it demonstrated that the efficiency of the machine itself was higher than we had estimated. The total result was according to estimate."[45]

Awards

Taylor was recognized for his brilliant work in his own time. In 1904, his paper "Ship-shaped Stream Forms" was awarded a gold medal from the British Institution of Naval Architects. "This was the first contribution for the future Admiral D.W. Taylor, U.S.N., and he looked forward to some future date 'when the naval architect, given the lines and speed of a ship, will be able to calculate the pressure and velocity of the water at every point of the immersed surface.' Rankine had discussed 'stream-line' surfaces in his paper of 1870 and had assumed flow to start and finish at 'foci,' now termed 'sources' and 'sinks' ahead and astern of a stationary body in a moving liquid. Taylor extended this conception and showed how the changes of pressure head could be calculated for two-dimensional flow around a stream flow… Taylor stated that the calculated increase of head at bow and stern corresponded fairly closely with the height of the bow wave."[46] This was the first time an American had received this award.

Taylor received a number of other awards including the Franklin Institute Gold Medal in 1907 and the John Fritz Gold Medal in 1931. Taylor remains among the least known of the Fritz medalists, who include President Herbert Hoover, Elmer Sperry, Orville Wright, Thomas Edison, George Westinghouse and Alexander Graham Bell. Taylor received the medal "…for outstanding achievement in marine architecture, for revolutionary results in persistent research in hull design, for improvements in many types of warships and for distinguished service as Chief Constructor of the United States Navy during the World War."[47] In 1918 he was voted into the National Academy of Science. Perhaps the greatest honor was paid to him in the form of the dedication of the new model basin at Carderock before his death. The new model basin is known as The David Taylor Model Basin.

Taylor and Froude

William Froude died in 1879, several years before David Taylor might have had the opportunity to meet him, but there is no doubt Taylor was familiar with the work of Froude and his son. Taylor wrote of the validity of Froude's Law of Comparison, "…The elder Froude made numerous experiments upon models, similar, but differing in size, and found that so far as careful observation could establish, the wave surfaces, and hence the stream lines, were similar at corresponding speeds. He found also that the Law of Comparison applied in such cases.

Froude recognized, however, that experiments with models, even though one were double the size of another, could hardly be considered conclusive in extending the law to full-sized ships.

Accordingly with the assistance of the Admiralty he carried out towing experiments to determine an actual curve of resistance for the *Greyhound*- a ship *172* feet long and of more than 1,000 tons displacement…

So far as I am aware the *Greyhound* experiments have not been repeated with any other ship, but the wave patterns of many men-of-war have been compared with those of their models at corresponding speeds by the Froudes, and appear to have been similar, so far as could be established by close observation.

All things considered, we appear fully justified in accepting the Law of Comparison."[48]

In addition, Taylor wrote to Sir Wescott Abell in 1933, "It has always seemed to me that William Froude was far ahead of his time, not only as a pioneer in the rolling of ships and resistance and propulsion, but as a genius who, with a model tank which was very crude compared with those of to-day, established methods and quantitative coefficients which served the naval architect for more than fifty years. We know now that Froude's coefficients can be improved upon, but for practical purposes, improvement has been astonishingly small."[49]

Conclusion

Taylor's 1900 report was colored by a renewed sense of urgency resulting from several events that took place as the EMB neared completion in 1899. The sinking of the USS *Maine* in Havana Harbor in January of 1898, the US Blockade of Cuba in support of the rebellion that April, the declaration of war on the United States by Spain shortly thereafter, and the Spanish-American War which followed had turned US Naval attention from the protection of her waterways and coasts to the protection of allies and possessions some distance away. To fulfill this obligation and realize the vision of the United stagnates as an active naval power in the global arena, more advanced ships were required; ships that could secure the United States' place as a major sea power. These ships would be built using the EMB to test their designs before construction began. The value and importance of model testing for the US was validated. It would play a role in the design of naval vessels from then on.

David Taylor's contributions to hydrodynamic theory and practice are respected to this day. His work not only established the validity of model testing for the United States but resulted in several theories that have stood the test of time:

Suction in Narrow Channel

Taylor's theory of the suction created when two large vessels pass in a narrow channel is respected to this day. Called as an expert witness, his testimony and presentation of his theory was an important factor in the outcome of *Olympic* vs. *Hawke*. His theory was further proven when, at a meeting of the Institution of Naval Architects in 1913, "some practical tests by Professor Gibson and J. H. Thompson were of considerable interest. A steam yacht of 88.5 ft. and 96 tons displacement was used to represent an *Olympic* and a 29.33 ft. motor boat of 2.6 tons represented a miniature *Hawke*, though the authors were careful not to offend the susceptibilities of either the Admiralty or the White Star Line by making any mention of the accident they were trying to reconstruct.

The slightest consideration of the streamline theory shows that there must always be suction between two vessels passing each other at all closely, as the level of water in the restricted channel between the two vessels will be less than that on the open-water sides. The only questions are the amount of the suction, the sheering effects set up and whether a small amount of helm can keep the smaller vessel out of danger.

The authors produced some quite satisfying collisions, and concluded that 'with vessels of the relative size used in these experiments, moving at speeds within 10 percent of each other, collision may be produced from a lateral distance as great as 3½ lengths of the smaller vessel, except in so far as prevented by helm action.'"[50]

The Speed and Power of Ships, in which the Taylor Series is laid out and the results explained, is regarded as a seminal work in the field of resistance.

The Bulbous Bow used on vessels worldwide today is a lasting tribute to the quality and importance of the work done by Taylor in the EMB.

Notes

1. Carlisle, R. P. (1998). *Where the fleet begins : a history of the David Taylor Research Center, 1898–1998*. Washington, Naval Historical Center: For sale by the U.S. G.P.O.
2. Hovgaard, W. (1941). "Biographical Memoir of David Watson Taylor, 1864–1940." *National Academy of Sciences of the United States of America Biographical Memoirs XXII* (Seventh Memoir).
3. Ibid.
4. Ibid.
5. Taylor, D. W. (1900). The United States Experimental Model Basin, Society of Naval Architects and Marine Engineers.
6. Ibid.
7. Ibid.
8. Ibid.
9. Ibid.
10. Ibid.
11. Ibid.
12. Ibid.
13. Ibid.
14. Ibid.
15. Ibid.
16. Ibid.
17. Ibid.
18. Ibid.
19. Ibid.
20. Ibid.
21. Ibid.
22. Ibid.
23. Ibid.
24. Ibid.
25. Ibid.
26. Carlisle, R. P. (1998). *Where the fleet begins : a history of the David Taylor Research Center, 1898–1998*. Washington, Naval Historical Center: For sale by the U.S. G.P.O.

27. Taylor, D. W. and E. A. Sperry (1909–1930). Taylor-Sperry Correspondence.
28. Carlisle, R. P. (1998). *Where the fleet begins : a history of the David Taylor Research Center, 1898–1998*. Washington, Naval Historical Center: For sale by the U.S. G.P.O.
29. Ibid.
30. Allison, D. K., Keppel, Ben G., Nowicke, C. Elizabeth *D. W. Taylor*, United States Government Printing Office.
31. Ibid.
32. Ibid.
33. Ibid.
34. Carlisle, R. P. (1998). *Where the fleet begins : a history of the David Taylor Research Center, 1898–1998*. Washington, Naval Historical Center: For sale by the U.S. G.P.O.
35. Ibid.
36. Allison, D. K., Keppel, Ben G., Nowicke, C. Elizabeth, *D. W. Taylor*, United States Government Printing Office.
37. Carlisle, R. P. (1998). *Where the fleet begins : a history of the David Taylor Research Center, 1898–1998*. Washington, Naval Historical Center: For sale by the U.S. G.P.O.
38. Taylor, D. W., United States. Maritime Commission. [from old catalog], et al. (1943). *The speed and power of ships; a manual of marine propulsion*. Washington,, U. S. Govt. Print. Off.
39. Carlisle, R. P. (1998). *Where the fleet begins : a history of the David Taylor Research Center, 1898–1998*. Washington, Naval Historical Center : For sale by the U.S. G.P.O.
40. Ibid.
41. Ibid.
42. Ibid.
43. Ibid.
44. Congress, L. o. (1920). The Wilbur and Orville Wright Papers. *General Correspondence: Taylor, David W., 1925*, Manuscript Division.
45. Ibid.
46. Barnaby, K. C. (1960). *The Institution of Naval Architects 1860–1960: An Historical Survey of the Institution's Transactions and Activities over 100 Years*. London, The Royal Institution of Naval Architects.
47. Allison, D. K., Keppel, Ben G., Nowicke, C. Elizabeth, *D. W. Taylor*, United States Government Printing Office.
48. Ibid.
49. Abbell, S. W. (1955). William Froude, M.A., LL.D., F.R.S. A Memoir. *The Papers of William Froude, M.A., LL.D., F.R.S. 1810–1879*. R. N. Duckworth, Captain A. D. London, The Instution of Naval Architects: xi-xiv.
50. Barnaby, K. C. (1960). *The Institution of Naval Architects 1860–1960: An Historical Survey of the Institution's Transactions and Activities over 100 Years*. London, The Royal Institution of Naval Architects.

Chapter 8
Early Aviators

> *Imagination is more important than knowledge. For knowledge*
> *is limited to all we now know and understand, while*
> *imagination embraces the entire world, and all there ever will*
> *be to know and understand.*
>
> Albert Einstein

Scientific advancement is a process in which the results of one theorist's work provide a starting point for those who follow. Froude's work unquestionably provided a starting point for Taylor. The Wright brothers provided a starting point for the next generation of aviation theorists and airplane manufacturers. It only stands to reason that there were people whose work contributed to the understanding of the Wrights. This work by the pioneers of aviation who preceded the Wrights provided the background for their historic flight.

Machinery

The ability to test a theory before throwing yourself into a life-threatening physical situation dependent upon the accuracy of that theory is key for any investigator wishing a long, fruitful life. Because of this, those wishing to test aerodynamic principles in action have long had a conundrum: How do they test their ideas for successful flight before leaping into the air?

At first, theorists focused their attention on the study of birds in flight. They observed the way they rose in the air, the way they remained in the air, and the way they attained forward motion while in the air. Those who tried taking to the skies by emulating what they'd observed quickly discovered that what appeared to be happening with the birds, was not actually happening the way they understood it to be. There had to be some significant errors in their analysis of bird behavior that led to repeated failed attempts at flight.

G. Hagler, *Modeling Ships and Space Craft: The Science and Art of Mastering the Oceans and Sky*, DOI 10.1007/978-1-4614-4596-8_8,
© Springer Science+Business Media, LLC 2013

This realization led to an investigation of the physical traits of birds that might aid in flight. Once theorists recognized the significance of the bird's hollow bones and specialized wing feathers, they were left to devise a new theory of flight. This new theory would eventually involve what are today known as the four forces of flight. But to attain this understanding, and put it into active use, early aviators needed to observe and understand the action of the air over their designs.

In the 1500s, Leonardo da Vinci had written in his Codex Atlanticus, "As it is to move the object against the motionless air so it is to move the air against the motionless object." He also wrote, "The same force as is made by the thing against air, is made by air against the thing."[1] These principles are fundamental to all modern wind tunnel work. In a wind tunnel, the air moves against the motionless object— the object being tested. The understanding that it was not necessary to move the object through the air would ultimately lead to better, more reliable data as well. But at the time, no such machines existed.

da Vinci also believed this force moving past the object was proportional to the surface area and the velocity of the body. ... he was right about the surface area but not about the velocity. In birds and aircraft, drag forces are proportional to the square of the velocity (Anderson 1997). That not only the surface area but also the shape of the body determines resistance to flow was also clear to da Vinci, witnessed by various drawings of streamlined bodies and projectiles based on the shapes of fish.[2]

If moving an object through the air and measuring the effect produced identical results to moving the air over the object while stationary according to what Giacomelli termed "the principle of aerodynamic reciprocity,"[3] it meant anyone wanting to get a taste of the behavior of an object in the air needed only a steady source of wind. The earliest accessible sources of steady airflow were the tops of hills with unobstructed airflow and the mouths of caves. These sites seem workable, and they often were, but the flow of air was neither predictable nor steady, leading to questionable results.

As men focused their attention on achieving heavier-than-air, powered flight, they knew they needed a more dependable way to gather vital information through research. Attention turned from using the wind as airflow for experiments, to generating airflow in the process of performing the experiment. The first attempts were made with a new type of machine, known as the *whirling arm*.

The Whirling Arm

A whirling arm is a device similar in function to a person spinning in one place with an object attached to a long string circling about him. The force generated is technically centrifugal, but in the process, air is flowing over the object at the end of the string. Whirling arms are far steadier and require far less energy on the part of the experimenter, than a simple string. The source of power is located at the center point—fulcrum—with an arm that extends from that fulcrum (Fig. 8.1). On some whirling arms, the arm that extends outward with a counterweight a brief distance from the fulcrum and opposite the portion of the arm used for the

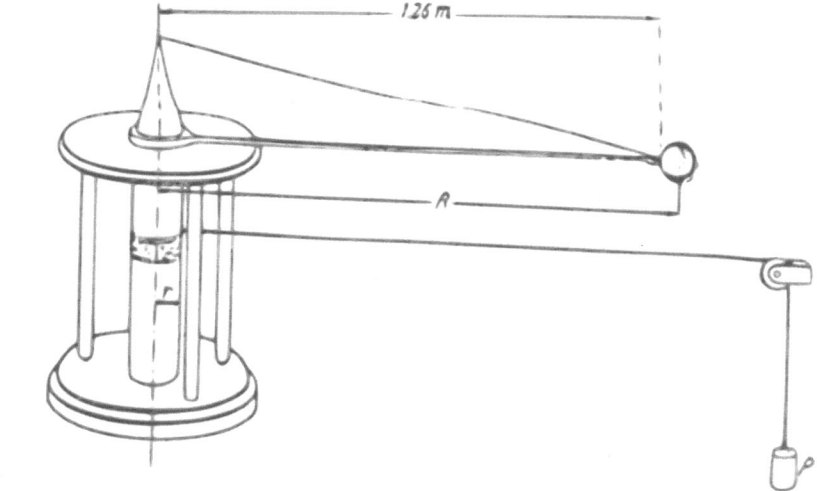

Fig. 8.1 Robin whirling arm. The whirling arm used by Robins

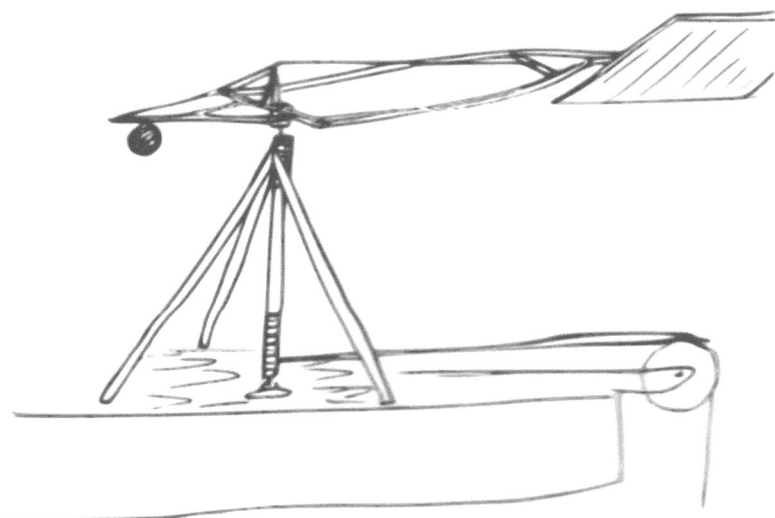

Fig. 8.2 Cayley whirling arm. The whirling arm used by Cayley

experiment (Fig. 8.2). An object is attached to the very end of the arm. Power is applied to spin the arm and readings are taken to further the investigator's understanding of the performance of an object with those characteristics moving through an airstream.

Benjamin Robins was the first to use a whirling arm in his work in 1742. Robins was an English mathematician and military engineer with an interest in ballistics. His used the whirling arm to test the manner in which shape affected the resistance experienced by an object moving through the air. By testing objects of the same material and total area—varying only in shape—through the same airstream, he was able to conclude that shape did affect drag.[4] His method did not allow for precision but it was more than adequate to advance the science of the day in a number of ways, attracting the attention of such notables as Leonhard Euler and the Royal Society in the process.

Robins proved that aerodynamic force does vary with the square of the relative velocity at speeds less than the speed of sound. He was also the first to observe the effect of the aspect ratio of a wing. He did this by proving that two bodies with the same frontal area but different shapes have different drag values.[5]

Robins' interest was in ballastics. He published the results of his work in a book entitled *New Principles of Gunnery Containing the Determination of the Force of Gunpowder and Investigations of the Difference in the Resisting Power of the Air to Swift and Slow Motions* in 1742. His paper, "Resistance of the Air and Experiments Relating to Air Resistance" appeared in the *Philosophical Transactions of the Royal Society of London* in 1746. His work was read by many researchers, including one of the greatest eighteenth century mathematicians and scientists, Leonhard Euler. Indeed, Euler was so excited about Robin's book that he personally translated it into German in 1745, adding some commentary of his own. In 1751, the year of Robins's death, it was translated into French.[6]

For his groundbreaking work, Robins was awarded the Copely Medal by the Royal Society in 1747. This was a significant degree of recognition by the scientific community in Britain, yet when Octave Chanute published his definitive survey on the technical development of the airplane to date, "Progress in Flying Machines," in 1894, he did not mention Robins or his work at all.[7] Robins may not have garnered his full degree of well-earned recognition, but his whirling arm had already forever changed the landscape for aerodynamic investigation.

John Smeaton was next to use the whirling arm. The year was 1759, when wind and water mills were major sources of power in Great Britain. Smeaton was a British civil engineer with an interest in the forces exerted on windmill blades by air and water. "For these experiments, Smeaton used Robins's invention of the whirling arm and adapted it such that the windmill blades not only translated in space via movement of the arm, but also such that the blades themselves rotated, thus simulating the actual operation of a windmill in the face of the wind. The windmill blades at the end of the whirling arm were spun by a cable and pulley mechanism activated by a falling weight."[8]

Smeaton published his results in a paper entitled "An Experimental Enquiry Concerning the Natural Powers of Water and Wind to Turn Mills, and Other Machines Depending on a Circular Motion" in the *Philosophical Transactions of the Royal Society of London, Volume 51, in 1759*. This paper included a table of aerodynamic force measurements on a flat surface perpendicular to the flow. The relationship expressed in this table was known as Smeaton's coefficient and had a value of 0.005. This value was of immense significance at the time, earning Smeaton a Copley Medal in 1759. The value would ultimately be proven incorrect, but it was taken as a given through the time of the Wright brothers. While used as a value of 0.005, Smeaton's coefficient introduced errors that hindered many attempts at flight.[9]

Sir George Cayley's whirling arm was next. He used his device to measure the drag and lift of a variety of airfoils. While Robins' whirling arm had an arm that was 4 feet long and spun by the action of a falling weight attached to a pulley and spindle arrangement with the tip reaching velocities of only a few feet per second, Cayley's whirling arm was 5 feet long and attained tip speeds between 10 and 20 feet per second. [10] The results from his tests led directly to the design of a successful hand-held glider. This small plane is believed to have been the first successful heavier-than-air vehicle in history, although a later glider by Penaud would have characteristics of greater longitudinal stability. In 1804 Cayley built and flew an unmanned glider with a wing area of 200 square feet. By 1852 he had a triplane glider design that incorporated many features of modern aircraft.[11] The results of his work led to the identification of the four forces of flight. An understanding of these forces would ultimately change the approach to flight and lead the way to the aircraft configuration we know today.

Otto Lilienthal, German engineer and the man who would become known internationally as the Glider King, was next to use a whirling arm for experimentation. Built in 1888, Lilienthal's whirling arm was 23-feet in diameter and 15 feet tall. He used his whirling arm to carry out aerodynamic tests on thousands of different airfoil shapes, arriving at a cambered shape as the most aerodynamically efficient. In 1891 Lilienthal used the results of his tests on cambered airfoils to successfully complete the first controlled, winged flight in a hang glider of his own design.

American-born Sir Hiram Maxim lived in England when he conducted his whirling arm tests. His device was 64 feet in diameter. "The arm boasted elaborate instrumentation to measure lift, drag, and relative air velocity."[12] He tested cambered airfoils with his whirling arm. Unhappy with the results, he became convinced that a wind tunnel would give him the desired precision.

In 1887, Samuel P. Langley gave whirling arm tests a try. At the time, Langley was the Secretary of the Smithsonian Institution and a widely respected scientist in the field of astronomy. His whirling arm was 60 feet in diameter and spun by a 10-horsepower engine. It could attain speeds of 100 mph. Langley was optimistic that manned flight could occur but unhappy with the problem inherent with the use of the whirling arm for aerodynamic testing.[13]

The Wind Tunnel

The inherent shortcomings of the whirling arm led to increasing levels of dissatisfaction with the testing process for experimenters. It also led to mounting unease over the accuracy of the results. One significant shortcoming of the whirling arm was that the air in the vicinity of the device had a tendency to spin in the direction of the device as the testing period wore on. This made it difficult, if not impossible, to accurately calculate the relative velocity between the component being tested and the moving air. Since this measurement provided essential information, the inability to count on their readings rendered the data obtained questionable at best. Another shortcoming of the whirling arm was that the device itself caused disruptions in the airstream. These disruptions in turn caused misleading results. When the results were being used to design machines intended to fly men through the sky, data accuracy was imperative. Clearly, a closed system with a controlled flow in one direction and nothing of the apparatus to interfere with the airstream was to be highly desired.

Francis Wenham tackled the task in the 1870s. A charter member of the Aeronautical Society of Great Britain, Wenham convinced the Society to raise the funds he needed to build a wind tunnel. Wenham designed the apparatus. It was built in 1871, the same year William Froude constructed his first model tank at Torquay, and Wenham was the first to use it.[14] Wenham's wind tunnel was 12 feet long and 18 inches square. Air was driven through a duct to the test section by a steam-powered fan. The model being tested was mounted in the test section and air traveled at a maximum speed of 40 mph. A number of specimens were tested. They included a variety of shapes, tested at various angles of attack.

It turned out that the airstream was not steady enough to provide precise results, but the results obtained were still of importance.[15]

Wenham's groundbreaking work demonstrated that the relationship of lift to drag of the test surfaces was higher than expected at low angles of attack. The high lift-to-drag ratios would result in wings that could support more substantial weights than anticipated. This made powered flight seem more attainable. The research also showed that long, narrow wings provided much more lift than stubby wings with the same area.[16] One sticking point to those who followed was the fact that Wenham tested only flat lifting surfaces in his wind tunnel.

Horatio Phillips built the second wind tunnel in the early 1800s. His tunnel was 6 feet long and 17 inches on each side. "He directed a jet of steam through the box, blasting a series of wing shapes that he placed inside the tunnel. He hoped to find out how fast the velocity of the oncoming airstream needed to be so that each different form, which carried equal weights, would remain suspended in the airflow."[17]

Phillips' tunnel overcame the airflow fluctuation problem experienced with Wenhams' tunnel by sucking the air through the entrance into the tunnel. It then went through a narrow area that reduced the flow area. The area was known as the throat and the model to be tested was mounted in there. The flow velocity was greater there and could reach about 41 mph.[18]

Phillips built his tunnel not only because he was dissatisfied with the airflow fluctuation in Wenham's tunnel but also because he wanted to test cambered (curved) airfoils as well. "Phillips designed a series of cambered airfoils with greater curvature over the top than on the bottom—so-called 'double-surface' airfoils... Phillips measured the aerodynamic performance of these airfoils in his wind tunnel and compared it with that of a flat plate also test in the same tunnel. The results were dramatic—*the cambered airfoils were considerably more efficient lifting shapes than a flat plate...* He properly recognized that when the flow moved over the curved upper surface of the airfoil the pressure decreased; hence the lifting action of the airfoil is due to a *combination* of the lower pressure exerted on the upper surface and the higher pressure exerted on the lower surface."[19]

The wind tunnel Hiram Maxim constructed at the end of the 1800s, when he became dissatisfied with the results from his whirling arm, was 12 feet long. The test section was three feet square. Maxim used two fans to blow air into the test section at 50 mph. The results from his tests with the whirling arm convinced Maxim that cambered airfoils provided the most lift with the least drag. His wind tunnel tests confirmed this. Maxim was also the first to understand that the total drag was more than the sum of the drag on each of the individual parts. This concept is known as "aerodynamic interference."[20]

By 1883, the Reynolds number demonstrated that the airflow pattern over a scale model would be the same for the full-scale vehicle if a certain flow parameter were the same in both cases.[21] This finding established the validity of wind tunnel testing.

Theory

British engineer John Smeaton started the process rolling in 1759 with the publication of a paper titled "An Experimental Enquiry Concerning the Natural Powers of Water and Wind to Turn Mills and Other Machines Depending on Circular Motion." The paper had nothing to do with flight in any form but it did address the relationship between pressure and velocity for objects moving in the air. Others would take Smeaton's work and derive what is still called Smeaton's coefficient. It is a constant of proportionality with the value of 0.005 and describes Smeaton's basic notion of pressure varying as the square of the velocity when applied to objects moving in air. Lilienthal and the Wright brothers would use this figure when making calculations for the lift of their aircraft. An error in the constant affected the outcome of these calculations and was eventually discovered and corrected.[22]

In 1799, Sir George Cayley had his concept for a fixed-wing aircraft stamped on a coin no larger than a US quarter (Fig. 8.3). The front of the coin showed an aircraft with a fixed wing, a fuselage occupied by a person, horizontal and vertical tails at the read end of the fuselage, and a pair of "flappers" for propulsion.[23] With the exception of the "flappers," Cayley's drawing was a prescient view of the functions behind the modern fixed-wing craft.

"His concept was further emphasized on the flip side of the disk, which showed, for the first time, a lift-and-drag diagram for a lifting surface. The arrow shows flow from right to left, and the heavy diagonal line represents a wing cross section at a rather large angle of attack to the flow. In the right triangle above the wing, we see the hypotenuse represents the resultant aerodynamic force, and the horizontal and vertical sides represent the drag and lift, respectively."[24]

Cayley, the father of aerodynamics, had successfully identified the four forces of flight. As a result, he "was responsible for the concept of the modern-configuration aircraft. He proposed a fixed wing to generate lift, a separate mode of propulsion to

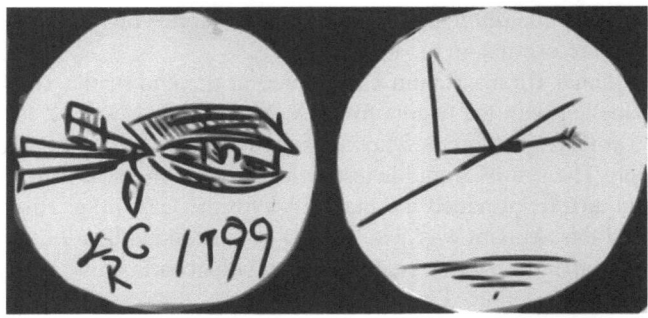

Fig. 8.3 Cayley coins. Sir George Cayley engraved the four forces of flight and his concept for a fixed-wing flying machine on a coin

overcome the 'resistance' (drag) to the machine's motion through the air, and both vertical and horizontal tail surfaces for directional and longitudinal static stability."[25]

Cayley wasn't content to leave it at that; he went on to test his fixed-wing concept by building a small, hand-launched glider. When tested in 1804 it successfully flew. Cayley went on to conduct many aerodynamic experiments with a whirling arm apparatus, publishing his results in a three-part paper entitled, "on Aerial Navigation," in 1809 and 1810. He turned his attention to balloon flight for a time, but in 1843 he again pursued his interest in heavier-than-air flight. From that point until his death in 1857, Cayley experimented with the angle of attack and the camber (curvature) of the airfoil.

As Cayley understood them, the forces were composed of two pair of opposing forces. They were *lift* and *gravity, thrust* and *resistance.* Only when all four were in balance could an object remain aloft and be in a position to attain forward motion. His vision was simple yet radical. In an aircraft of his design, there would be more than one mechanism at work to produce the four forces of flight. Identifying the forces made it possible to test components to determine the ideal configuration of each for the intended purpose.

But first Cayley tested his fixed-wing concept by building a small, hand-launched glider (Fig. 8.4). "The wing was essentially a kite fixed to the wooden rodlike fuselage and was inclined 6 degrees to the rod by a small peg at the leading edge of the wing. The tail could be set to any angle with the rod. A small weight dangled from the nose to adjust the location of the center of gravity. When Cayley successfully flew this hand-launched glider in 1804, it became the first modern configuration airplane in history to fly."[26]

Cayley wanted information about the variation of lift with the angle of attack of a fixed wing. No such data existed; so Cayley built and operated a whirling arm device, making Cayley the first person to use a whirling arm for aeronautical purposes. He was also among the first to notice a possible error in Smeaton's coefficient. "His investigations included measurements of the aerodynamic drag on a flat plate oriented perpendicular to the airstream. His results gave a value of 0.0037 for Smeaton's coefficient, in contrast to the value of 0.005 published by Smeaton. Today

Fig. 8.4 Cayley glider. Cayley's design for a glider using his fixed-wing design

we know that a proper value of Smeaton's coefficient is 0.003; Cayley's measurements were getting closer to the truth."[27]

Cayley published the results of his aerodynamic experiments with the whirling arm device in a three-part paper entitled, "on Aerial Navigation," in 1809 and 1810. He turned his attention to balloon flight for a time, but in 1843 he again pursued his interest in heavier-than-air flight. From that point until his death in 1857, Cayley experimented with the angle of attack and the camber (curvature) of the airfoil.

As a direct result of Cayley's work, those interested in flight now understood that taking off from the ground under their own power was simply not possible. There was no way to generate enough lift to overcome the force of gravity on the mass of the human body. They also understood that they although they could take off from a high point, no amount of flapping was going to allow them to overcome the force of gravity or generate sufficient thrust to move them forward. The best man would be able to do was glide to the ground.

In 1923, French aviation historian Charles Dollfus wrote of Cayley: "The aeroplane is a British invention: it was conceived in all essentials by George Cayley, the great English engineer who worked in the first half of the last century. The name of Cayley is little known, even in his own country, and there are very few who know the work of this admirable man, the greatest genius of aviation. A study of his publications fills one with absolute admiration both for his inventiveness, and for is logic and common sense. This great engineer, during the Second Empire, did in fact not only invent the aeroplane entire, as it now exits, but he realized that the problem of aviation had to be divided between theoretical research—Cayley made the first aerodynamic experiments for aeronautical purposes—and practical tests, equally in the case of the glider as of the powered aeroplane."[28]

It was not unusual for those working on the problems of powered flight to be unaware of the work of others. This was certainly the case with Alphonse Penaud and George Cayley. In 1871, Penaud built a small model airplane that was powered by twisted strands of rubber (Fig. 8.5).

Penaud called his craft the *planophore*, flying it successfully at Tulieries Gardens in Paris on August 18, 1871. The Wrights received one of these gliders as a gift from their father when they were boys. They referred to it as "the bat" and tried unsuccessfully to build a replica once the original was destroyed.

Fig. 8.5 The Penaud Planophore

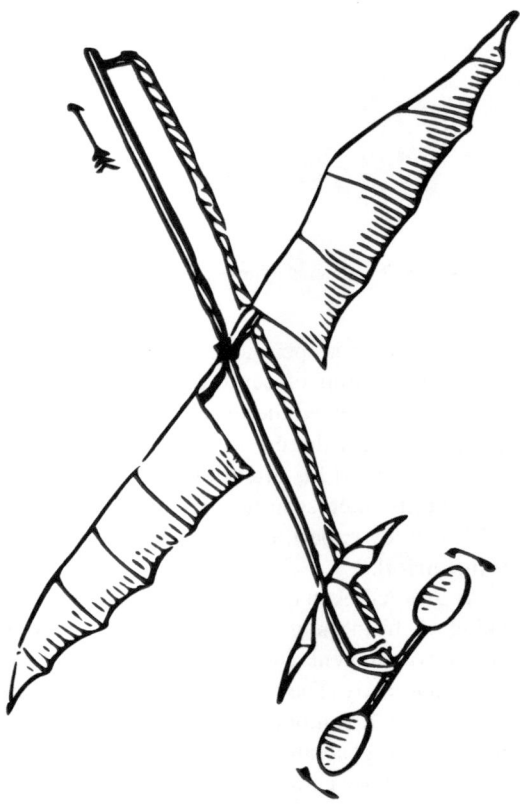

Penaud was unaware of Cayley's glider of 1804. Even if he had known of it, however, Penaud took a different approach to the balance and stability of the planophore in flight. While Cayley had inclined the tail with the front edge upward, Penaud set his horizontal tail at a negative incidence of −8 degrees relative to the chord line of the wing. Although Penaud could not have known it at the time, it would prove to be the case that a negative tail for a rear-mounted tail is necessary for the longitudinal balance of an airplane.[29]

Penaud's contribution to his chosen field grew from his understanding of the theory and the practice of stability. Cayley had located the wing on his glider at an extreme forward location with the center of gravity *behind* the center of pressure of the wing because of the positive tail setting. This resulted in a longitudinally unbalanced airplane. Penaud, with his negatively pitched tail, located his wing along the fuselage. Its center of pressure was *behind the center of gravity* of the whole machine. Penaud did this because he recognized the lift of the wing had to act behind the center of gravity if the planophore was to have longitudinal stability. The nose would pitch down with this arrangement, but the negative tail setting angle resulted in a *downward* lift on the tail, pitching the tail up and balancing the air-

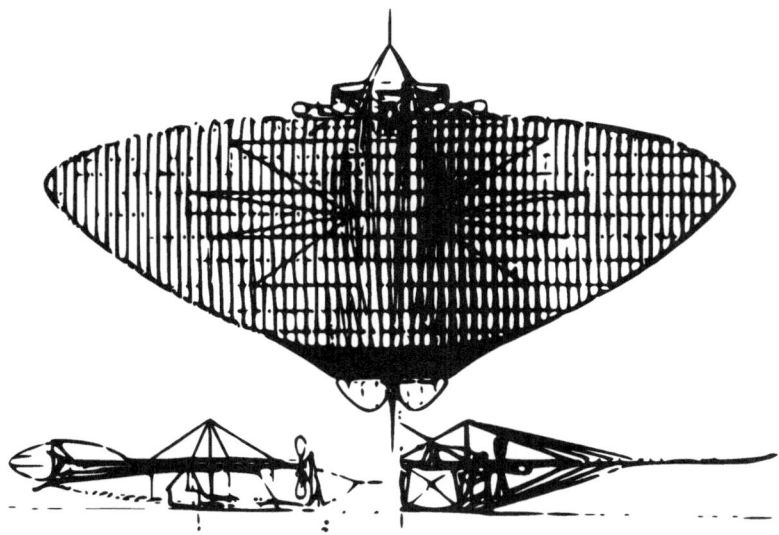

Fig. 8.6 Penaud large glider. Penaud's design for a full-sized glider

plane. Placing the wing and pitching the tail as he did ensured the airplane would restore equilibrium when the nose of the airplane experienced a gust of wind and the nose pitched upward, increasing the wing lift momentarily due to the increased angle of attack. The net effect would be for the nose to pitch back down, restoring equilibrium.[30]

"An observer accurately explained this behavior of Penaud's planophore in an account given to the 1874 annual meeting of the Aeronautical Society of Great Britain: "By the alternate action of the weight in front and the rudder [horizontal tail] behind the plane [wing], the equilibrium is maintained. The machine during flight, owing to the above causes, describes a series of ascents and descents after the manner of a sparrow." "[31]

With the success of his planophore, Penaud was encouraged to design a large, full-size flying machine. He received a patent for his design in 1876. Penaud's design (Fig. 8.6) was for "a two-seat monoplane with two tractor propellers (propellers mounted ahead of the wind and oriented to *pull* the airplane through the air, in contrast to pusher propellers mounted behind the wing that *push* the airplane through the air). The propellers rotated in opposite directions to cancel the torque effect of each. Probably for aesthetic appeal, the wings had an elliptical planform. ... The airfoil sections of Penaud's wing were cambered... The wings had a small dihedral angle of two degrees for lateral stability. The machine had two elevators at the rear and a fixed vertical fin with an attached vertical rudder. The cockpit had a glass dome, a single control column to operate the elevators and rudder, and instruments such as a compass, a level, and a barometer (for measuring altitude). Penaud included retractable landing gear with shock absorbers, and a tail skid. The machine also sported pontoons because Penaud felt that any full-scale flight experiments should be conducted over water."[32]

Penaud was unable to find funding for his magnificent machine. His design was met with derision by a public that believed only madmen would take to the skies in a heavier-than-air craft. Penaud committed suicide in 1880, but his work became widely known. It continued to influence other inventors into the twentieth century.

Horatio Phillips is not remembered only because he was the man who built the second wind tunnel in history. He is remembered because he is the man who definitively proved that *lift* was not generated only according to Newton's Third Law. "Although George Cayley had also alluded to this fact [that the lifting action of the airfoil is due to a *combination* of the lower pressure exerted on the upper surface and the higher pressure exerted on the lower surface], the prevailing intuition throughout most of the 19th century was that the lifting action of an inclined plane moving through the air was due to the 'impact' of air on the lower surface—a mental picture wrongly reinforced by the Newtonian flow model. Phillips, by designing double-surface airfoils, with more curvature over the top than the bottom, was qualitatively trying to encourage the formation of low pressure on the top surface, hence taking advantage of this fact to obtain a more efficient airfoil. Phillips's results were widely disseminated, and all serious fling machine developers after him used cambered airfoils."[33]

The Wright brothers would go on to perform their own airfoil tests in a wind tunnel of their own design and construction. Orville would state in a deposition given in 1921:

"Cambered surfaces were used prior to our experiments. However, the earlier experimenters had so little accurate knowledge concerning the properties of cambered surfaces that they used cambered surfaces of great inefficiency, and the tables of air pressures which they possessed concerning cambered surfaces were so erroneous as to entirely misled them. They did not even know that the center of pressure traveled backward on cambered surfaces at small angles of incidence, but assumed that it traveled forward. I believe we possessed in 1902 more data on cambered surfaces, a hundred times over, than all of our predecessors put together."[34]

Orville Wright may not have been entirely generous about the state of knowledge before the brothers performed their wind tunnel tests, but he was most likely correct about the accuracy and amount of data they had amassed on cambered surfaces through their own research by 1902.[35]

Octave Chanute had retired from a successful engineering business when he turned his attention to the new science of aviation in 1890. Chanute set out to gather all the data he could from those involved in flight experimentation. He amassed data from around the world and, with the publication of his book, "Progress in Flying Machines," in 1894 he became the international authority on flight. Chanute was in his 60s at the time, and did not personally fly his gliders, but he did design gliders that were piloted by others.

Convinced that the way to achieve greater lift was the addition of wings. This idea, for stacking wings above one another, had been proposed by Wenham in 1866. Lilienthal had successfully flown such a glider in the 1890s. For his attempts,

Chanute used a "strut-wire" brace similar to those used in his bridge-building work. The design was successful and the Wright brothers referred to it as the Chanute "double-decker." They based their glider designs on the Chanute's.

Chanute had some success with his gliders, but this transplanted Frenchman's vital role was the dissemination of knowledge related to the burgeoning field of aviation at the end of the nineteenth century. He had an active correspondence with aviation innovators in Europe and the United States. He corresponded with everyone from Pilcher to the Wright brothers, provided encouragement and served as a sounding board to the Wright brothers, and was one of the few who were welcome at Kitty Hawk. Chanute delighted in promoting the free exchange of information between those involved in the problem of flight. He was quick to advise those he corresponded with to patent their work, and there was some strain in his relationship with the Wright brothers over questions of his importance to their work in the years after the successful flight and Kitty Hawk and until Chanute's death in 1910. Even with all that, it is difficult to describe his influence in the Wrights' career as anything less than formidable.

Chanute's interest in an open discussion of aviation research and findings seems largely to have been in the interests of the day when flight would be a reality. In 1894 he wrote, "So may it be; let us hope that the advent of a successful flying machine, now only dimly foreseen and nevertheless thought to be possible, will bring nothing but good into the world; that it shall abridge distance, make all parts of the globe accessible, bring men into closer relation with each other, advance civilization, and hasten the promised era in which there shall be nothing but peace and goodwill among all men."[36]

Otto Lilienthal was another prominent forerunner or the Wright brothers. Lilienthal was already a noted German engineer when his book, "Der Vogelflug als Grundlage der Fliegekunst," that analyzed birds' wings and applied his findings to mechanical flight was published in 1889. Translated into English, *Birdflight as the Basis of Aviation* was a definitive source of aerodynamic information. The data in the book was the result of thousands of Lilienthal's own tests made on a variety of airfoil shapes attached to a whirling arm[37] (Fig. 8.7).

One significant contribution of Lilienthal's was the use of coefficients to express his findings. "Rather than reporting just the raw data, the actual values of the aerodynamic force, Lilienthal divided his measured forces at various angles of attack by the force measured when the wing was at a 90-degree angle of attack—when the wing was perpendicular to the flow. The *ratios* were dimensionless values, called *force coefficients*, which vary with angle of attack. In so doing, Lilienthal's coefficients are not compromised by any uncertainty in Smeaton's coefficient and are not a function of velocity. When the aerodynamic force is reported in coefficient form, the influence of Smeaton's coefficient and velocity are simply divided out. In modern aerodynamics today, we generally deal with aerodynamic lift and drag in terms of *lift and drag coefficients* exclusively. The origin of this modern use can be traced to Lilienthal. Some of Lilienthal's data were published and disseminated widely in the form of a table of coefficients,

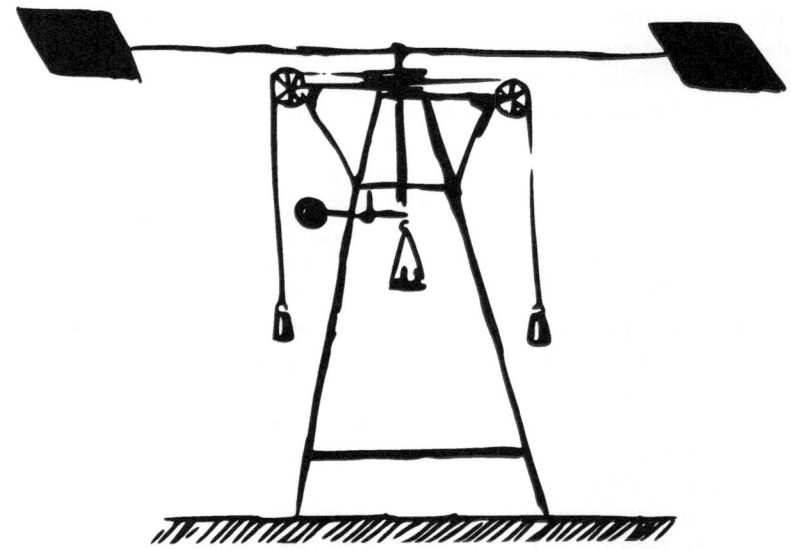

Fig. 8.7 Lilienthal whirling arm. The whirling arm used by Lilienthal

called "the Lilienthal tables" by the Wright brothers and others. The Lilienthal tables became famous."[38]

But research alone was not enough for the future Glider King. Lilienthal wrote, "One can get a proper insight into the practice of flying only by actual flying experiments. The journey in the air without the use of the balloon s absolutely necessary in order to gain a judgment as to the actual requirements for an independent flight. It is in the air itself that we have to develop our knowledge of the stability of flight so that a safe and sure passage through the air may be obtained, and that one can finally land without destroying the apparatus."[39]

His childhood attempts to strap on wings and take off had failed, but his resolve remained strong. Lilienthal built several types of mono-wing gliders that were designed to distribute the weight of the craft and of Lilienthal as evenly as possible. The gliders for which he was known around the world were based on his own findings. He used a cambered wing that was attached to his body at the shoulders. He built a hill specifically for the purpose of take off. This hill permitted him to take off into the wind, no matter which direction the wind was blowing from. He hung beneath it in a vertical position and controlled the glider by shifting his weight, but his vertical position gave him limited stability (Fig. 8.8). In all, he made more than 2,000 flights. The Wright brothers, along with scores of flight enthusiasts, followed news of Lilienthal's flights. On August 8, 1896, his glider went into a stall and crashed. His physician recalled Lilienthal's injury: "I can still see him today, lying on his back with his beautiful, full, blond beard, not remarking about any pain. I basically did not take his injury very seriously, as he could still move both arms well, though he was completely paralyzed from the

Fig. 8.8 Lilienthal glider. Otto Lilienthal, The Glider King, and his glider. *Source*: Library of Congress Prints and Photographs Division Washington, DC 20540, USA

waist down, a sure sign that his spine must be broken."[40] Lilienthal died 2 days after his accident.

But Lilienthal had found a way to stay aloft. By flying into the wind, he generated lift as the oncoming wind blew over the airfoil of his wing and the conditions for the Bernoulli Principle came into play. Especially with his curved wing models, the wind blowing up and over the top of the wing moved more quickly than the wind moving beneath the wing. This created an area of low pressure above the wing, causing the craft to rise as the air beneath the wing exerted pressure in an upward direction. However, Lilienthal did not have a mechanism for generating thrust once the momentum of his jump dissipated. He also had no means of stability other than shifting his weight to move the center of gravity of his craft. Since he hung below his glider in a vertical position rather than in a horizontal position, this gave him control over the angle of attack in a limited way. As a result, he did not have the ability to recover from a stall caused by something as simple as a gust of wind.

Lilienthal's influence on the Wrights was immense. It was upon his untimely death that an active interest in flight awakened in the brothers. They shared Lilienthal's conviction that the only way to master flight was to spend time in the air. It was not an activity that could be perfected by watching from the ground. They also understood the process of learning as Lilienthal described it, "The manner in which we have to meet the irregularities of the wind when soaring in the air can only be learned by being in the air itself. At the same time it must be considered that one single blast of wind can destroy the apparatus and even the life of the person flying.

This danger can only be avoided by becoming acquainted with the wind by constant and regular practice, and by perfecting the apparatus so that we may achieve safe flight. The only way which leads us to a quick development in human flight is a systematic and energetic practice in actual flying experiments."[41]

Wilbur first gave Lilienthal public credit in his address to the Western Society of Engineers in 1901 when he stated, "Lilienthal not only thought, but acted; and in so doing probably made the greatest contribution to the solution of the flying problem that has ever been made by any one man. He demonstrated the feasibility of actual practice in the air, without which success is impossible."[42]

Wilbur last publicly praised Lilienthal shortly before his own death in 1912, in a piece that was published posthumously. "Of all the men who attacked the flying problem in the 19th century, Otto Lilienthal was easily the most important. His greatness appeared in every phase of the problem. No on equaled him in power to draw new recruits to the cause; no one equaled him in fullness and dearness of understanding of the principle of flight; no one did so much to convince the world of the advantages of curved wing surfaces; and no one did so much to transfer the problem of human flight to the open air where it belonged. As a scientific investigator none of his contemporaries was his equal."[43]

Samuel Pierpont Langley was next to attempt powered, heavier-than-air flight. Langley had not yet been appointed Secretary of the Smithsonian when he turned his attention to aviation in 1886. His reputation had been largely made for his groundbreaking work with studies of the sun and sun spots. However, he was able to construct and operate a major facility for the sole use of obtaining aerodynamic data on the Alleghany Observatory grounds in Pittsburgh. Funding came from a wealthy friend and by 1887 Langley's whirling arm was in action.

Langley conducted his research over a 4-year period. At the end of that time, he published a book, *Experiments in Aerodynamics*, that was the first substantive American contribution to aerodynamics. Langley had read the work of Wenham and Phillips at the start of his whirling arm experiments, but was unaware of Lilienthal's experiments. As a result, Langley felt himself entering an area in need of accurate data. His published data was all about flat plates, although he later examined cambered surfaces. His aerodromes, in fact, used cambered wings.[44]

The most controversial conclusion reached by Langley on the basis of his experimentation is the "Langley Law." It simply states that the power required for a vehicle to fly through the air *decreases* as the velocity increases.[45] "This conclusion flies in the face of intuition, which is why Langley labeled it as 'paradoxical.' It was considered to be misleading at best by some contemporaries and outright wrong by others. Lilienthal and the Wright brothers rejected this conclusion outright. In a meeting of the British Association for the Advancement of Science at Oxford in August 1894, Langley presented a short paper summarizing his work and conclusions; he was criticized and taken to task by both Lord Kelvin and Lord Rayleigh— formidable opposition to say the least. Indeed, Langley has been derided for this power law to the present day.

Langley's conclusion, however, was based on his experimental data, and those data *consistently* supported it. ... Examining Langley's data we note that they were

Fig. 8.9 Langley houseboat. Langley's Aerodrome A atop the houseboat used for launching.
Source: NASA/courtesy of nasaimages.org

all taken at velocities of 20 meters per second or less…had his whirling arm allowed testing at velocities greater than 22 meters per second, Langley would have noted a reversal in this data trend, and most likely the Langley power law would never have existed."[46] Unfortunately for Langley, the bumps to his reputation were just beginning.

In 1887, Langley was appointed Secretary to the Smithsonian Institution. It had become clear to him that to convince the rest of the world that mechanical flight was possible with the engines then available, he would have to do more than conduct laboratory experiments. He would need to build a successful airplane. As a result, the man viewed by many as the most prestigious scientist in the United States, set out to build a successful flying machine. Some of Langley's first attempts were with rubber-powered model airplanes he called aerodromes. The purpose of experiments with those craft was to determine the practical conditions of equilibrium and of horizontal flight. Unfortunately, the results of this work were mixed and not very instructive.[47] Langley abandoned these attempts and turned his attention to steam-powered flight of full-sized aerodromes.

To accomplish this he built the first full-sized aerodrome—his name for his flying machines. In the interests of safety, Langley chose to launch his aerodromes off a houseboat on the Potomac by means of a catapult (Fig. 8.9). On the surface, this may strike some as ludicrous, but the likelihood of death in the event of a failed attempt weighed ever heavy on the minds of aviation pioneers. Langley explained, "As the end of the year 1892 approached and with it the completion of an aerodrome of large size which had to be started upon its flight in some way, the method and place of launching it pressed for decision. One thing at least seemed clear. In the present stage of experiment, it was desirable that the aerodrome—if it must fall—fall into water where it should suffer little injury and be readily recovered, rather than anywhere on land, where it would almost certainly be badly damaged."[48] It was believed that a successful landing on the Potomac would require rebuilding the aerodrome, but an unsuccessful landing on the Potomac would not likely end in the death of the man onboard.

It took 3 years of failure before Langley achieved success, but on May 6, 1896 he achieved the first ever, successful flight of an engine-powered, heavier-than-air flying machine. The attempt was quite public. In fact, Alexander Graham Bell was on hand and wrote in a letter describing the successful flights he'd observed, "it seemed to me that no one could have witnessed these experiments without being convinced that the possibility of mechanical flight had been demonstrated." [49]

It wasn't long before Langley was ready for a new challenge. He wrote in a letter to Octave Chanute in 1897, "If anyone were to put at my disposal the considerable amount—fifty thousand dollars or more—for... an aerodrome carrying man or men, with a capacity for some hours of flight, I feel that I could build it and should enjoy the task." He went on to predict that he could accomplish this feat within two or three years from the time he would start.[50]

With those words Langley opened the door to what would be the devastating end to his successful career. He acquired the requisite $50,000 from the War Department and constructed his aerodromes at four times the scale of his success-ful gliders. Unfortunately, Langley assumed the increase in scale would not impact the performance of the craft. Not only did the up-scaled aerodromes fail to perform as intended, they failed utterly. And publicly.

By once again basing his operation on the houseboat on the Potomac, it was virtu-ally guaranteed there would be plenty of people to observe Langley's progress. There would also be plenty of people on hand for Langley's latest attempts at flight. The first took place on October 7, 1903. The press was on hand to watch as, with much fanfare, the cord to the catapult was cut. As soon as the cut was made, the aerodrome "tumbled over the edge of the houseboat and disappeared in the river, sixteen feet below. It simply slid into the water like a handful of mortar..."[51] That failure was attributed to a problem with the launching mechanism. The second attempt took place on December 8, 1903. The wheels suffered a total collapse this time and the aerodrome again fell into the river. Fortunately for Manly, Langley's assistant and the man chosen to be at the controls on the two machines Langley attempted to fly, crashes involving the water proved not to be fatal. Manly was unhurt but Langley's attempts were over. Over that is but for the ridicule Langley withstood until his death on February 27, 1906. Considering Langley's reputation for brilliant work in the field of astronomy, his position as the third Secretary of the Smithsonian Institution, and the general level of respect for the man and his work until the time of his ventures in aviation, it was a decidedly sad ending to an otherwise illustrious life.

The Wright's had relatively nothing to do with Langley and little interest in his experiments. Even when Langley asked to visit them at Kitty Hawk, they politely declined. For the Wright brothers it was time to take their turn at heavier-than-air flight. They had the benefit of Cayley's work with the four forces of flight; Lilienthal's philosophy, figures, and experience with his gliders; Smeaton's coefficient; and Octave Chanute as a sounding board. They also had the accumulated experimentation and experience of many other theorists and adventurers at their disposal.

Sorting through it all, determining which parts were reliable and which were not would require all the meticulous attention to detail the Wrights could muster. They understood that the ability to control the angle of attack—the position of the craft in relation to the wind—was key to surviving their attempts at flight. They believed as Lilienthal had, that the more time spent flying, the more likely they would be to conquer the problems of controlled, manned, powered flight. They were determined to amass a significant amount of flying time in the safest manner possible.

If Orville and Wilbur Wright lacked money, time was even harder to come by due to the demands of their bicycle shop. They did not have long periods of uninterrupted time to spend arriving at the configuration of their craft. When confronted with results that were inconclusive, they wasted no time in generating their own data through the use of scale model tests in the tradition of naval architects such as William Froude and David Taylor. Their efforts would result in the successful flight at Kitty Hawk on December 17, 1903. With this flight, the Wrights would prove to believers and naysayers alike that it was possible for humans to successfully employ Cayley's four forces of flight and join the birds in the sky.

Notes

1. Videler, J. J. (2006). *Avian Flight*, Oxford University Press.
2. Ibid.
3. Anderson, J. D. (1997). *A history of aerodynamics and its impact on flying machines*. Cambridge; New York, Cambridge University Press.
4. "Centenniel of Flight."
5. Anderson, J. D. (2002). *The airplane: a history of its technology*. Reston, VA, American Institute of Aeronautics and Astronautics.
6. Ibid.
7. Ibid.
8. Ibid.
9. Ibid.
10. "Centenniel of Flight."
11. Ibid.
12. Baals, D. D. and W. R. Corliss (1981). *Wind tunnels of NASA*. Washington, D.C., Scientific and Technical Information Branch for sale by the Supt. of Docs., U.S. G.P.O.
13. Ibid.
14. "Centenniel of Flight."
15. Ibid.
16. Ibid.
17. Ibid.
18. Ibid.

19. Anderson, J. D. (2002). *The airplane: a history of its technology.* Reston, VA, American Institute of Aeronautics and Astronautics.
20. "Centenniel of Flight."
21. Ibid.
22. Ibid.
23. Anderson, J. D. (2002). *The airplane: a history of its technology.* Reston, VA, American Institute of Aeronautics and Astronautics.
24. Anderson, J. D. (1997). *A history of aerodynamics and its impact on flying machines.* Cambridge; New York, Cambridge University Press.
25. Ibid.
26. Anderson, J. D. (2002). *The airplane: a history of its technology.* Reston, VA, American Institute of Aeronautics and Astronautics.
27. Ibid.
28. Ibid.
29. Anderson, J. D. (2004). *Inventing flight: the Wright brothers & their predecessors.* Baltimore, Johns Hopkins University Press.
30. Ibid.
31. Ibid.
32. Ibid.
33. Anderson, J. D. (2002). *The airplane: a history of its technology.* Reston, VA, American Institute of Aeronautics and Astronautics.
34. Ibid.
35. Ibid.
36. Chanute, O. (1894). *Progress in Flying Machines,* The American Railroad Journal.
37. Anderson, J. D. (2002). *The airplane: a history of its technology.* Reston, VA, American Institute of Aeronautics and Astronautics.
38. Ibid.
39. Lilienthal, O. (1896). Flying As A Sport. *The American Magazine,* Frank Leslie Publishing House: 3.
40. Anderson, J. D. (2002). *The airplane: a history of its technology.* Reston, VA, American Institute of Aeronautics and Astronautics.
41. Lilienthal, O. (1896). Flying As A Sport. *The American Magazine,* Frank Leslie Publishing House: 3.
42. (1903). Annual Report of the Board of Regents of the Smithsonian Institution Showing The Operations, Expenditures, and Condition of the Institution for the Year Ending June 30, 1902. *Doc. No. 484, Part 1.*
43. Anderson, J. D. (2002). *The airplane: a history of its technology.* Reston, VA, American Institute of Aeronautics and Astronautics.
44. Ibid.
45. Ibid.
46. Ibid.
47. Ibid.
48. Ibid.
49. Ibid.
50. Ibid.
51. Ibid.

Chapter 9
The Wright Brothers

The desire to fly is an idea handed down to us by our ancestors who, in their grueling travels across trackless lands in prehistoric times, looked enviously on the birds soaring freely through space, at full speed, above all obstacles, on the infinite highway of the air.

Wilbur Wright

It's a common knowledge that Wilbur and Orville Wright were the first to successfully fly a piloted, powered, heavier than air object. The process leading to that successful flight in Kitty Hawk, North Carolina, was fraught with obstacles ranging from what the brothers believed to be faulty data to difficulties in obtaining the ideal materials. Through it all, the Wrights combined a systematic approach with dogged determination to overcome each impediment. The net result for these first aeronautical engineers was not only that first successful flight on December 17, 1903. The Wrights also proved the value of testing airfoil models in a wind tunnel to anticipate the performance of a variety of wing structures during the design phase and before construction of an actual aircraft began.

There were already generations of seaworthy ships to serve as examples of what did and did not work when the time came to improve upon the performance of ocean going vessels in the late nineteenth century. Even with that volume of hands-on experience, scale model testing quickly earned a permanent place in the development cycle because of the inarguable benefits it brought to the design process. It's no wonder then in the case of piloted flying machines, with scant history as a guide, that model testing would prove useful early on.

G. Hagler, *Modeling Ships and Space Craft: The Science and Art of Mastering the Oceans and Sky*, DOI 10.1007/978-1-4614-4596-8_9,
© Springer Science+Business Media, LLC 2013

The Wright Approach

As a starting point for their experiments with flying machines, the Wrights could only look to craft that had come close to success but had failed. They were very aware of the untimely demise of the pilots and/or inventors who were aloft when each craft failed to perform as anticipated. As a result, the Wright brothers were not just faced with the problem of improving on what had come before. They were faced with the problem of surviving the process of improving on what had come before.

There had been many advances in the field of hydrodynamics by the time the brothers began their experiments with flight. However, there was a great division between academicians involved in the pursuit of scientific knowledge and the craftsmen involved in the pursuit of successful flight. Because of this, the Wrights and others of their time found themselves ignorant of much of the work that had been done and might have been of use to them.[1] Given the brothers analytical approach, it was simply a matter of time before the methodical testing of components to form a basis for their experiments occurred to them.

An Interest in Flight Awakens

The Wrights had an interest in aviation that dated back to their boyhood. They wrote, "It was in the autumn of 1878 that our father brought us home one evening a toy, which he held half hidden in his hand for a while to excite our curiosity and then tossed it into the air. Instead of falling to the ground, as we expected it would, the thing soared in the air, across to the other side of the room, where it struck against the opposite wall and then finally fluttered lightly to the floor. It was a little toy called scientifically the helicopter, but we boys promptly christened it 'bat.' It was a miniature airship of bamboo and cork covered with paper and having two little propellers turned by twisted rubber bands. So fragile a plaything could not last long in the hands of children, but the impression it made upon our minds remained, permanently."[2] (Fig. 9.1).

The possibilities of flight excited by their experience with "bat" remained in the background as the brothers pursued other interests and it was not until 1896, and the "deplorable death" of Lilienthal, that Wilbur turned his impressive intellectual curiosity to the matter of flight. In his 1901 address to the Western Society of Engineers, titled "Some Aeronautical Experiments," he would share, "My own active interest in aeronautical problems dates back to the death of Lilienthal in 1896. The brief notice of his death which appeared in the telegraphic news at that time aroused a passive interest which had existed from my childhood and led me to take down from the shelves of our home library a book on Animal Mechanism, by Professor Marey, which I had already read several times. From there," Wilbur continued, "I was led to read more modern works, and as my brother soon became

Fig. 9.1 Their father brought
a Penaud Planophore home
for the boys

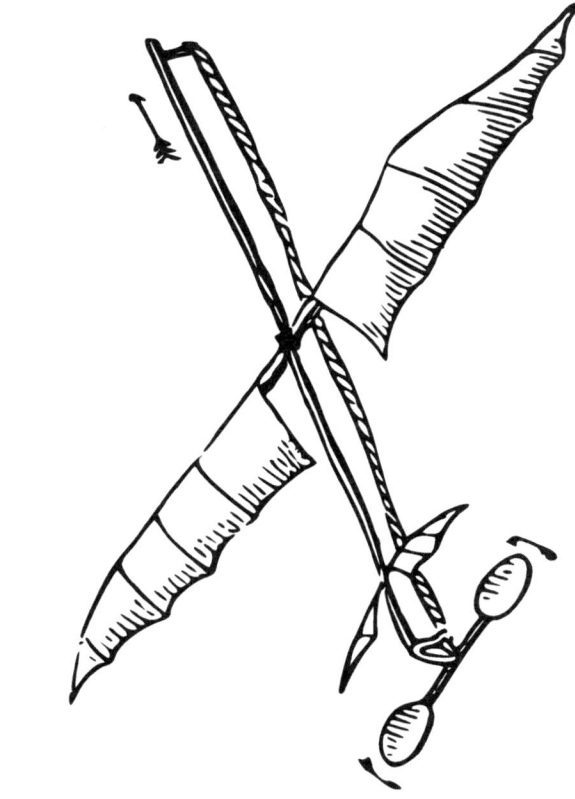

equally interested with myself, we soon passed from the reading to the thinking, and finally to the working stage."[3]

By May 30, 1899, Wilbur declared himself "an enthusiast, but not a crank…" in a letter to the Smithsonian Institution in which he requested any materials pertaining to research on flight that had been done to date. He wrote, "I am about to begin a systematic study of the subject in preparation for practical work to which I expect to devote what time I can spare from my regular business. I wish to obtain such papers as the Smithsonian Institution has published on this subject, and if possible a list of other works in print in the English language. I am an enthusiast, but not a crank in the sense that I have some pet theories as to the proper construction of a flying machine. I wish to avail myself of all that is already known and then if possible add my mite to help on the future worker who will attain final success."[4]

Wilbur may have written the letter requesting the materials but Orville was familiar with the works as well, as is evident from his response to questions during a deposition he gave during a trial in 1920. "… On reading the different works on the subject we were much impressed with the great number of people who had given thought to it, among these some of the greatest minds the world has produced.

But we found that the experiments of one after another had failed. Among those who had worked on the problem I must mention Leonardo Vinci, one of the greatest artists and engineers of all time; Sir George Cayley, who was among the first of all inventors of the internal-combustion engine; Sir Hiram Maxim, inventor of the Maxim rapid-fire gun; Parsons, the inventor of the turbine steam engine; Alexander Graham Bell, inventor of the telephone; Horatio Phillips, a well-known English engineer; Otto Lilienthal, the inventor of instruments used in navigation and a well-known engineer; Thomas A. Edison; and Dr. S. P. Langley, Secretary and head of the Smithsonian Institution."[5]

The failure of those who had come before them notwithstanding, their reading left the brothers both intrigued with the possibilities of gliding as a sport and determined to escape the fate of Lilienthal and Pilcher. Wilbur recalled, "We found that both these experimenters had attempted to maintain balance merely by the shifting of the weight of their bodies. Chanute, and I believe all the other experimenters before 1900, used this same method of maintaining the equilibrium in gliding flight. We at once set to work to devise a more efficient means of maintaining equilibrium."[6]

This idea of the pilot as an active participant was integral to the Wrights' view from the start. Also from the start, the Wrights viewed their glider as a system made up of components designed for identifiable tasks. Specifically, they needed components capable of creating sufficient lift, generating ample thrust, and exerting control over the four aerodynamic forces of flight: lift, drag, thrust, and weight.[7] Controlling *pitch* and *yaw* had already been examined. That left the problem of *lateral control*. The brothers understood that, "a rolling motion, hence, lateral control, could be obtained by simultaneously setting the right wing at one angle of attack to the flow and the left wing at another angle of attack, such that the different lift forces on the two wings would induce a rolling motion."[8] The problem then was not what to do but how to do it.

Wilbur soon came up with a practical means of lateral control, what the brothers would call "wing-warping." He shared his findings with Orville who recalled, "... He demonstrated the method by means of a small pasteboard box, which had two of the opposite ends removed. By holding the top forward corner and the rear lower corner of one end of the box between his thumb and forefinger and the rear upper corner and the lower forward corner of the other end of the box in like manner, and by pressing the corners together the upper and lower surface of the box were given a helicoidal twist, presenting the top and bottom surfaces of the box at different angles on the right and left sides. From this it was apparent that the wings of a machine of the Canute double-deck type, with the fore-and-aft trussing removed, could be warped in like manner, so that in flying the wings on the right and left sides could be warped so as to present their surfaces to the air at different angles of incidence and thus secure unequal lifts on the two sides ..."[9]

To test their theory of "wing-warping," the brothers built a biplane kite, fully expecting that their methodology, in which the trailing edge of the wings of a glider were bent in opposite directions to reduce roll, would resolve one of the greatest impediments to sustained flight. When the kite was a success, it fueled their desire to build and test a piloted flyer. They wrote to the US Weather Bureau for information on the best possible sites to conduct their experiments. Included in their requirements were low population density, sustained winds, and a reasonably dry season, warm enough for outdoor work during the winter months that made up the off-season for their bicycle business. Very mindful of the damage that could be done to the pilot and equipment when a flying machine of any type crashed, they also inquired about areas meeting the former criteria and having a large area of sand as well. Kitty Hawk was chosen for its remote location, prevailing winds, and large expanse of sand, thought to be less punishing to pilot and craft in the event of a crash than grass or rock.

And So They Begin

Despite the inherent danger, the brothers were intent both on building and personally piloting their craft because Wilbur was certain, as he said in his presentation to the Western Society of Engineers, "The person who merely watches the flight of a bird gathers the impression that the bird has nothing to think of but the flapping of its wings. As a matter of fact this is a very small part of its mental labor. To even mention all the things the bird must constantly keep in mind in order to fly securely through the air would take a considerable part of the evening. If I take this piece of paper, and after placing it parallel with the ground, quickly let it fall, it will not settle steadily down as a staid, sensible piece of paper ought to do, but it insists on contravening every recognized rule of decorum, turning over and darting hither and thither in the most erratic manner, much after the style of an untrained horse. Yet this is the style of steed that men must learn to manage before flying can become an everyday sport. The bird has learned this art of equilibrium, and learned it so thoroughly that its skill is not apparent to our sight. We only learn to appreciate it when we try to imitate it."[10]

When Wilbur alluded to men learning to manage this style of steed, he introduced the novel concept of man doing more than shifting his weight to control the path of his glider, as had been the practice up until that time. Just what it was the pilot would do was still unknown, but Wilbur felt the only way to know was to actively experience flight and to amass enough flying time to form a significant body of knowledge about control issues during flight. In his address he likened learning about flight and its requirements to learning to ride a difficult horse. "...there are two ways of learning how to ride a fractious horse: One is to get on him and learn by actual practice how each motion and trick may be best met; the other is to sit on a fence and watch the beast a while, and then retire to the house and at leisure figure out the best way of overcoming his jumps and kicks. The latter system is the safest, but the former, on the whole, turns out the larger proportion of good riders. It is very much the same in learning to ride a flying machine; if you are looking for perfect safety, you will do well to sit on a fence and watch the birds; but if you really wish to learn, you must mount a machine and become acquainted with its tricks by actual trial."[11]

On May 13, 1900, Wilbur wrote the first of many letters in what would become a 10-year correspondence with Octave Chanute, then considered the international authority on flight due to the success of his bi-wing glider tests and the publication of his book, "Progress in Flying Machines." Chanute was in correspondence with inventors and craftsmen around the world and served not only as a valuable source of information but also as a sounding board for Wilbur as the brothers pursued their ultimate goal of piloted flight.

Wilbur began that initial letter with a description of his interest in flight. "For some years I have been afflicted with the belief that flight is possible to man. My disease has increased in severity and I feel that it will soon cost me an increased amount of money if not my life..."[12] Since the purpose of his letter was to engage Chanute in a discussion of the current state of research he continued, "with this general statement of my principles and belief I will proceed to describe the plan and apparatus it is my intention to test. In explaining these, my object is to learn to what extent similar plans have been tested and found to be failures, and also to obtain such suggestions as your great knowledge and experience might enable you to give me."[13]

It is clear that Wilbur had given a great deal of thought to the subject of flight from the time of Lilienthal's death in 1896 to that spring day in 1900. Also clear was Wilbur's initial belief that secrecy was not a serious consideration, as he wrote to Chanute, "I make no secret of my plans for the reason that I believe no financial profit will accrue to the inventor of the first flying machine, and that only those who are willing to give as well as to receive suggestions can hope to link their names with the honor of its discovery. The problem is too great for one man alone and unaided."[14]

Wilbur's plan was that he would, "in a suitable locality erect a light tower about one hundred and fifty feet high. A rope passing over a pulley at the top will serve as a sort of a kite string. It will be so counter balanced that when the rope is drawn out one hundred & fifty feet it will sustain a pull equal to the weight of the operator and apparatus or nearly so. The wind will blow the machine out from the base of the tower and the weight will be sustained partly by the upward pull of the rope and partly by the lift of the wind. The counter-balance will be so arranged that the full decreases as the line becomes shorter and ceases entirely when its length has been decreased to one hundred feet. The aim will be to eventually practice in a wind capable of sustaining the operator at a height equal to the top of the tower. The pull of the rope will take the place of a motor in counteracting drift. I see, of course, that the pull of the rope will introduce complications which are not met in free flight, but if the plan will only enable me to remain in the air for practice by the hour instead of by the second, I hope to acquire skill sufficient to overcome both these difficulties and those inherent in flight. Knowledge and skill in handling the machine are absolute essentials to flight and it is impossible to obtain them without extensive practice."[15]

Wilbur never built his tower, practicing with first gliders as kites (Fig. 9.2) and then as piloted at Kitty Hawk. However, his tentative plan does highlight his deep concern over safety and the strength of his conviction that they must pilot their

Fig. 9.2 Glider as kite. 1901 glider being flown as a kite. *Source*: Library of Congress Prints and Photographs Division Washington, DC 20540, USA

own craft because a lack of flying time/experience was the root cause of earlier misfortune. He estimated that with a total of nearly 2,000 flights, Lilienthal had amassed only about 5 h total flying time. Wilbur's goal was to achieve that many hours in a matter of days or weeks. He was certain that the only way to overcome the failures that went with lack of flight time was for the brothers to build and pilot their own machines.

Wilbur was also convinced that identifiable obstacles blocked the path to successful flying. In his paper for the Western Society of Engineers in 1901 he wrote that they were, "of three general classes: (1) Those which relate to the construction of the sustaining wings; (2) those which relate to the generation and application of the power required to drive the machine through the air; (3) those relating to the balancing and steering of the machine after it is actually in flight.... This inability to balance and steer still confronts students of the flying problem, although nearly eight years have passed. When this one feature has been worked out, the age of flying machines will have arrived, for all other difficulties are of minor importance."[16]

Lilienthal's Data and the Need for Model Testing

The Wrights initially concluded that part of the problem of balance and steering on the craft flown by Lilienthal, Pilcher, and others to date was due to the fact that the pilot was perpendicular to the craft, creating drag in much the same way a cyclist sitting completely upright interfered with the forward progress of a bicycle. In fact, the early gliders were all much like current day hang gliders in that respect, with a stationary wing mounted at the shoulders and the pilot suspended below by some type of harness. Wilbur and Orville planned that the pilot on their gliders would be horizontal, reducing the drag by being atop the glider itself. With that as a plan they set to work, first on gliders and later on a flyer.

To determine the size of the wings, the Wrights used two figures generated by others. One was the Smeaton Coefficient (k) discussed in Chap. 4. This coefficient was widely accepted at the time, in error, as having a value of .005, although there were also those who were publicly questioning the accuracy of the figure. This measurement was fundamental to many of the other measurements that would be made because it was used universally as a reference point. The other figures the Wrights would use were those generated by Lilienthal over the course of his work. These *lift coefficients* (cl) from Lilienthal's tables were essential to their calculations as well because they helped determine the optimal size of the wings. (See more about Lilienthal's methodology in Chap. 4.) With the Smeaton Coefficient and Lilienthal's lift coefficients to guide them, the Wrights calculated that a 50 pound, two-wing craft carrying a 150-pound pilot and traveling at about 20 mph would need each wing to be 5 feet by 20 feet. Having no reason to question the results obtained from their calculations, the Wrights' first glider, the 1900 glider, was constructed to these specifications with wings of this size.

After initial excitement at seeing their glider rise successfully aloft as a kite, the brothers quickly grew frustrated. The 1900 glider never attained adequate lift to carry a pilot. To generate increased lift, the brothers increased the *chord* of the wing on the 1901 glider, doubling the weight of the craft in the process, yet they still achieved only about one-third of the lift anticipated by Lilienthal's data. It was enough to carry a pilot, however, and the brothers were able to make several gliding flights.

These poor lift results could easily have led the Wrights to question their approach to flight. They did not, primarily because the brothers made measurements of every variable available to them each time they flew their glider/kite and these measurements served to give them confidence in the figures they generated. They wisely used multiple approaches to verify their figures with one significant set of measurements made while flying parallel to the side of a hill. By measuring the angle of the hill, the Wrights could determine the ratio of lift to drag. With this figure, they could predict the number they'd find in Lilienthal's data. Their numbers consistently disagreed with Lilienthal's and so the brothers gradually came to question the accuracy of Lilienthal's tables as well as Smeaton's coefficient.

Questioning Lilienthal's numbers meant questioning the existing aerodynamic science of the day. It also meant questioning the work of Lilienthal, a man both brothers credited with doing more to advance the science of flight than any other individual of their time. As the result of experimentation done between the end of the 1901 flying season and the start of the 1902 flying season, the brothers produce tables of data of their own to replace the data in Lilienthal's tables. Their work would also add to the proof that Smeaton's coefficient of .0054 should be actually have been about .00033. They couldn't have known achieved either of these milestones without the confidence and conviction to undertake a careful investigation of their own.

By the end of the 1901 flying season, despite having solved the problem they referred to as "well-digging," the tendency of the glider to spin out of control in yaw when the pilot tried to bank and turn, the brothers were very discouraged. Their new, larger wings were still not living up to expectations. In fact, they generated only one-third of the lift anticipated with the use of Lilienthal's tables. The brothers were on the verge of giving up altogether when they returned home on August 22. Wilbur recalled "On the train ride home … when we looked at the time and money which we had expended, and considered the progress made and the distance yet to go, we considered our experiments a failure. At this time I made the prediction that men would sometime fly, but that it would not be within our lifetime."[17]

Any serious thoughts of abandoning their work were put aside a few days after arriving home when Wilbur received an invitation from Chanute to speak at the Western Society of Engineers on September 18. Wilbur's initial reaction was to decline the invitation because he wasn't sure they'd done anything worth speaking about. It was his sister who convinced him to accept the invitation. He went on to prepare a speech, "Some Aeronautical Experiments," based on a thorough reexamination of everything they had done to date. When the speech was printed and available to the public, it quickly became the acknowledged expert source of aeronautical knowledge. The preparation of the paper and the examination of their work to date would be the final step that spurred the brothers to undertake their own formal model testing.

Testing Begins

Wilbur felt a great deal of anxiety about his September presentation in Chicago. He wrote to Chanute, "… I make no pretense of being a public speaker." In response to Chanute's asking how he felt about making the presentation a ladies night, Wilbur wrote, "As to the presence of ladies, it is not my province to dictate, moreover I will already be as badly scared as it is possible for man to be, so that the presence of ladies will make little difference to me, provided I am not expected to appear in full dress, &c."[18]

Wilbur's presentation included a concise yet thorough walk through the experiments and gliding flights he and Orville had conducted through August 1901. He described their initial interest in flight, the impact of Lilienthal's work and untimely death, and a detailed explanation of their methodology and results. He included their suspicion that Lilienthal's figures might be in error. He also stated that they thought Smeaton's coefficient might be off by as much as 20%, not as controversial as questioning Lilienthal because Langley and others had already published work in which they called Smeaton's coefficient into doubt. Because of Wilbur's initial reluctance to present their findings, as well as the short timeframe between the invitation and the actual presentation, it was agreed that modifications could be made to the final version of "Some Aeronautical Experiments." For the Wright brothers, this meant a brief opportunity to run some additional tests to verify their suspicions about Lilienthal's figures before turning their attention back to their cycling business.

While Wilbur, dressed splendidly in Orville's formal clothing, was in Chicago presenting his paper, Orville constructed a wind tunnel for the tests they planned. The brothers used this early model to test their first versions of the equipment they would perfect for the second, improved version of the wind tunnel that soon replaced it. It was in this improved wind tunnel that they would conduct the airfoil tests that formed the basis of their future work. But before any work was performed in a wind tunnel, the brothers made an initial, simple test of the correctness of Lilienthal's figures using an instrument that was wholly familiar to them; the bicycle. These results would show that Lilienthal's numbers were definitely incorrect. The results would be included in the revised manuscript included in the Transactions of the Western Society of Engineers.

In keeping with Wilbur's methodical nature, he wrote about the upcoming bicycle test to Chanute on September 26, 1901. "I am arranging to make a positive test of the correctness of the Lilienthal coefficients at from 4°–7° in the following manner. I will mount a Lilienthal curve 1 sq. ft. and a flat plane of .66 sq. ft. on a bicycle wheel in the position shown. The vision is from above. The distance from the centers of pressure to centre of wheel will be the same for both curve and planed. According to Lilienthal tables the 1 sq. ft. curve at 5° will just about balance the .66 sq ft. plane at 90°. If I find that it really does so no question will remain in my mind that these tables are correct. If the curve fails to balance the plane I will cut down the size of the plane till they do balance. I hope to make the test on the first suitable day. If you have any suggestions to make regarding it, or any error in the principle employed to point out, I should be very glad to know it.

As soon as I have made this test I will revise my manuscript and forward it at once."[19]

To perform the described test, the Wrights balanced a curved surface atop a plane surface attached to a bicycle. They placed this on a wheel that was mounted horizontal to the handlebars and attached to the handlebars, above the front wheel. To power the wheel that was mounted at the handlebars, the Wright's rode the bicycle through the streets of Dayton. Ever thorough, they first rode at right angles to the wind in one direction, then in the other, on a calm day, noting and comparing the results. With a difference of only about 2 degrees, due to direction, they next replaced the curved surface with a plane surface and repeated the tests.

According to Lilienthal's tables, "the model wing set at a five-degree angle of attack would generate enough lift to balance the flat plate exactly... The model-wing surface required an angle of attack of *18 degrees* to balance the plate, more than three times what the table indicated it should be. The Wrights were able to conclude definitively from this that either Lilienthal's lift coefficients or the Smeaton value was in error."[20]

With the tests concluded, Wilbur followed up with a letter to Chanute on October 6 in which he wrote, "we have made the experiment of balancing a curved surface against a plane surface 66 percent as large, placed normal to the wind, and find that instead of 5° as called for in Lilienthal's table an angel of 18° was required." The letter continues: "The results obtained, with the rough apparatus used, were so interesting in their nature, and gave evidence of such possibility of exactness in measuring the value of $P_{(tang.\ a)}/P_{90}$, that we decided to construct an apparatus specifically for making tables giving the value of $P_{(tang.\ a)}/P_{90}$ at all angles to 30° and for surfaces of different curvatures and different relative lengths & breadths. The new apparatus is almost as simple to construction as the vane already used and the values given are lifts in percentages of P_{90} without extended calculations."[21]

The new apparatus mentioned would be the wind tunnel. It would be used to generate a new set of figures to guide the design of their flying machines. The Wright brothers were not the first to use a wind tunnel, but they were the first to run a series of tests on airfoils they created for the purpose and then to compare their findings to data collected during actual flights for the purpose of making changes to the design of their craft. In fact, the work they did in 1902 and 1903 would take advantage of the first use of models in the aircraft design process. Their wind tunnel work would be as seminal to the design of aircraft as Froude's work in his model basin had been to the design of vessels. Their work would lay the groundwork for scientific aeronautical model testing for those who followed.

Wind Tunnel Tests

Given the Wrights penchant for using whatever scraps they had on hand when embarking on a new project, the first wind tunnel they constructed was a crude rectangular machine. Before long they realized the results of their tests could be compared to their own flight data and used to determine the optimal size for the wings under a variety of conditions. To facilitate this work, the brothers decided they would build an improved version of their wind tunnel and use it to test for lift and drag as variables on airfoils that simulated a variety of wing shapes.

"The wind tunnel was operating by mid-October... The flow duct was 6 feet long with a square cross section 16 inches on each side. There was a glass window on top for observing the tests. The airflow was driven by a fan powered by the central power plant of the Wrights' bicycle shop—a 1-horsepower gasoline engine connected to the fan via shafts and belt drives. The maximum velocity attainable in the wind tunnel was about 30 miles per hour. The tunnel was housed on the second floor of the bicycle shop, where all of the testing took place..."[22]

Making certain the air flowed through the wind tunnel without creating turbulence required some ingenuity on the part of the Wrights. It took several months to perfect but by using a sort of honeycomb arrangement at the intake of the tunnel, they were able to create a smooth flow of air for their purposes.[23] Because the brothers mistakenly installed the tunnel's two-bladed fan upstream, their laboratory itself was the return path for the air rushing out of the tunnel test section.[24] Because of this, it was essential to the necessary flow of air was that no other movement take place on the second floor while the tests were being conducted. To ensure this, Orville would observe through the glass on the top of the tunnel while Wilbur conducted the active part of the process. Orville could not move about the area because any movement by him caused a distortion in the readings.

The Balances

The balances were odd-looking devices made of old hacksaw blades and discarded bicycle spokes. They didn't look very sophisticated but, "In reality, they were marvels of simplicity and sophistication. Originally, the Wrights tried to measure both lift and drag with a single instrument." Those results were, "subject to errors of perhaps ten percent," as Wilbur wrote to Chanute.[25] As a result, they decided to use separate balances for lift and drag.

"The lift balance was in theory very similar to the earlier bicycle apparatus and the balancing vane used in the first tunnel... The lift of the model wing surface was measured in terms of an opposing force exerted on a flat plate oriented perpendicular to the airstream. In other words, the lift force generated by the curved surface would be measured as a fraction of the oppositely directed force resulting from the wind hitting the flat plate. Put even more simply, they were expressing lift as the ratio of these two forces. This meant that the pressure on the flat pate would serve as the common standard against which the lifts of the various wing shapes tested would be measured."[26]

The balances used for the final version of the wind tunnel not only gave more accurate readings than earlier versions; they enabled the Wrights to calculate the lift coefficient directly, without the need to use Smeaton's coefficient. With the new balances they could also set the airfoil to a variety of angles of attack while the flat plate was mounted on equipment that was free to circle around a fixed point. "During the initial trials the plate created a disturbance of the flow in the tunnel, so the brothers replaced it with four narrow strips of an equivalent area. The fingerlike attachments added to the ungainly look of the device. With the wind turned on, the ratio of the force of lift generated by the wing to *both* the force exerted on the plate plus the drag force produced by the wing was indicated by an angle traced out by a pointed connected to the arms carrying the surface and the four strips that replaced the plate. But at this stage the Wrights wanted to record only the lift of the model wing. To eliminate the effect of the drag component of the total aerodynamic force acting on the wing, a mechanical readjustment of the arms of the balance was required. After resetting the balance, a second reading was taken. The angle now traced out by the pointer only indicated the component of aerodynamic force generated by the wing owing purely to lift."

"Constructing the balance in such a way as to factor out the effect of drag mechanically was one of the most remarkable features of the instrument's design. Even from a modern perspective, this aspect of the balance was an incredibly impressive piece of engineering."[27]

The final step in the process of determining lift required the use of basic geometry to take the sine of the angle to arrive at the coefficient of lift. A closer look at their methodology reveals a deep understanding of geometry and trigonometry. Both Wilbur and Orville were strong mathematicians, but it was Orville who had proven himself a strong mathematician while a student. Knowing this, we can conclude that Orville played a key role in the analysis of the airfoil tests.[28]

"Like their instrument for measuring lift, the Wrights' drift balance was cleverly designed to measure the aerodynamic forces acting on a wing in terms that could be substituted directly into the lift and drag equations they were using to predict the performance of their aircraft. The drift balance again showed the uncommon ability of the Wrights to develop practical mechanical devices that mirrored their conceptual analysis of a problem."[29]

Airfoils

The *airfoils* themselves were made from scraps of 20-gauge steel. These models of wing designs were cut, hammered, and soldered into a variety of different shapes depending upon their purpose. Some were designed to test the *aspect ratio*. For this purpose the brothers used flat plates that had the same area but differing aspect ratios. Others were intended to test the *camber* of a wing. On these, the aspect ratio and area were constant but the curvature of the airfoil differed. Still others were *parabolic foils*. The area and aspect ratio were constant. The camber varied and the foil itself had curved sides.[30]

Each airfoil was between three and nine inches long and had a piece of metal about ¾″ high and ¾″ long soldered on the back of the airfoil, perpendicular to the surface. These raised pieces of metal would be inserted into two parallel pieces of metal that were soldered to the balances used inside the actual tunnel, to hold the airfoil securely in place during testing. This ingenious pincer and tab arrangement allowed for the airfoils to be held in precise positions without impeding the flow of air in any way.

The brothers tested over 200 different airfoils in late October and early November of 1901 as they perfected their balances. When they were done, they felt confident that their methodology was sound and would result in reliable data. On November 22, the brothers were ready to begin the formal testing of a smaller sample of airfoils. For these tests they made thirty-eight airfoils. "They tested camber ratios from 1/6 to 1/20; the location of maximum camber ranged from near the leading edge to the midchord position. The planform shapes included squares, rectangles, ellipses, surfaces with raked tips, and circular arc segments for leading and trailing edges meeting at sharp points at the tip. They also examined tandem wing configurations (after Langley's aerodromes), biplanes, and triplanes. Finally, Wilbur and Orville had to end these experiments because of the press of business.... These experiments, conducted over less than an two-month period, produced the most definitive and practical aerodynamic data on wings and airfoil obtained to that date. They gave the Wrights proper aerodynamic information on which to design a proper flying machine."[31]

When the Wrights were satisfied with the results for a particular *camber*, they used that angle in the shape of their glider wing. The data from the testing was then compared to the results achieved from a test of the wing. This iterative process was the basis for design improvements made by the Wrights.

Wilbur later wrote, "It is difficult to underestimate the value of that very laborious work we did over that homemade wind tunnel. It was, in fact, the first wind tunnel in which small models of wings were tested and their lifting properties accurately noted. From all the data that Orville and I accumulated into tables, an accurate and reliable wing could finally be built. Even modern wind tunnel data with the most sophisticated equipment varies comparatively little from what we first discovered. In fact, the accurate wind tunnel data we developed was so important, it is doubtful if anyone would have ever developed a flyable wing without first developing this data. Sometimes the non-glamorous lab work is absolutely crucial to the success of a project."[32]

The Wrights formally recorded the results of 43 surfaces and multiwing forms tested on their lift balance, and another 48 on their drift balance. They evaluated the surfaces from 0 to 45 degrees angle of attack. When running the determinations, they took great care to maintain consistency in their procedure and to ensure that no outside influences would adversely affect the readings.[33]

"The heart of any successful wind tunnel is its balance system-the apparatus that measures the aerodynamic forces acting on the model. The Wrights built two balances—one for lift and a second for drag. The balances never measured actual forces; they simply compared test airfoils with reference airfoils or other forces on calibrated flat surfaces. This approach allowed the Wrights to rapidly pit one airfoil against another and select the best from many configurations."[34]

Wilbur and Orville concluded their testing in mid-December of 1901 because they needed to turn their attention back to their cycling business. Even when Chanute offered the possibility of funding that would have allowed them to continue their work with the airfoils, the brothers politely declined.[35] During their testing period, the Wrights investigated not only lift and rag but also aspect ratio and the effect of a thick versus sharp leading edge and the impact of varying the gap between biplane and triplane wing arrangements.[36]

The enduring legacy from the Wrights' experimentation was the tables they created from their experimentation. "One table gives lift coefficient tabulated versus angle of attack α. Another table gives the drag-to-lift ratio as a function of α. These tables supplanted the Lilienthal table in all respects. At the time, they represented the most valuable technical data in the history of applied aerodynamics."[37]

Back to Kitty Hawk

With their new tables in hand, the Wright brothers were anxious to return to their camp at Kitty Hawk (Fig. 9.3). Before the 1901 flying season, with no way of fore-seeing the legal difficulties that would come about after their successful flight in 1903, Wilbur had written Chanute, "we note what you say in regard to the discretion and reliability of Messrs Huffaker & Spratt. We have felt no uneasiness on this point, as we do not think the class of people who are interested in aeronautics would naturally be of a character to act unfairly. The labors of others have been of great benefit to us in obtaining our understanding of the subject and have been suggestive and stimulating. We would be pleased if our labors would be of similar benefit to others. We or course would not wish our ideas and methods appropriated bodily, but if our work suggests ideas to others which they can work out on a different line and reach better results than we do, we will try hard not to feel jealous or that we have been robbed in any way. On the other hand we do not expect to appropriate the ideas of others in any unfair way, but it would be strange indeed if we should be long in the company of other investigators without receiving suggestions which we could work out in such a way as to further our work."[38]

Fig. 9.3 Kitty Hawk. The Wright's work shed at Kitty Hawk, North Carolina. *Source*: Library of Congress Prints and Photographs Division Washington, DC 20540, USA

With tables in hand a year later, the brothers appear to have been slightly more guarded with their work. When asked whether or not they wanted to publish their work, Wilbur deferred to Chanute and did not pursue the matter himself. As a result, the Wrights were the only ones with the data they possessed at the time of their return to Kitty Hawk in 1902 and 1903.

Armed with reliable data, Wilbur and Orville soon made modifications to their craft that enabled it to carry a pilot aloft and in charge at significant heights and for significant periods of time. By 1903 they were ready to tackle the problems of propulsion and with their successful flight on December 17, they proved to themselves and to the world that it was possible to control the aerodynamic forces in play during the piloted flight of a heavier than air machine generating its own thrust.

After Success

As soon as the brothers wired a message home to tell their family of their success, the wire was leaked to the press in Norfolk, touching off a number of incorrect and inaccurate reports. Wilbur wrote to Chanute, filling him in on the stiff winds and the various attempts they'd made on that important day but he left out any specific technical details. Soon after, Wilbur issued a statement to the press, explaining that the brothers owed nothing to any one, for their success as they'd financed their work themselves. Chanute replied with a letter on the January 14, 1904 in which he wrote, "In the clipping which you sent me you say: "all the experiments have been conducted at our own expense, without assistance from any individual or institution." Please write me just what you had in your mind concerning myself when you framed that sentence in that way."[39]

Wilbur replied on the 18th, "The object of the statement, concerning which you have made inquiry, was to make it clear that we stood on quite different ground from Prof. Langley, and were entirely justified in refusing to make our discoveries public property at this time. We had paid the freight, and had a right to do as we pleased. The use of the word 'any,' which you underscored, grew out of the fact that we found from articles in both foreign and American papers, and even in correspondence, that there was a somewhat general impression that our Kitty Hawk experiments had not been carried on at our own expense etc. We thought it might save embarrassment to correct this promptly."[40]

Wilbur continued his correspondence with Chanute until Chanute's death in 1910. Wilbur was involved in speaking and writing about their work until his own death from typhoid fever on May 12, 1912. After Wilbur's death, Orville continued in the world of aeronautics, building a laboratory and working on inventions as he chose. He was active in promoting flight and speaking about the historic flight he and his brother had made. He was a charter member of NACA, the National Advisory Committee for Aeronautics, executive committee. During his work with NACA, he corresponded with Rear Admiral David Taylor on the matter of the 1903 propellers. The NACA was involved in virtually all areas of aeronautics. The 12 unpaid members consulted the federal government on several aviation-related issues during the first decade, including recommending the inauguration of airmail service and studying the feasibility of flying the mail at night.[41] Orville was a part of NACA until his death on January 30, 1948, a total time of nearly 30 years.

Conclusion

The Wright brothers began their pursuit of flight as enthusiastic amateurs. Their belief that Lilienthal's contention that the only way to master flight was to get in the air and fly, along with their meticulous nature and innovative spirit carried them through the difficult tasks required to design and test a variety of craft until the right combination of wing type and angle was discovered. They carried out their work at their own expense, consulting frequently with Octave Chanute as they progressed. Along the way, they established the legitimacy of wind tunnel testing for airfoils and wing models, added their own experimentation to the correct value of Smeaton's coefficient, generated data in the form of coefficients, and replaced Lilienthal's tables with figures generated by their own investigations.

By the time they flew that December day at Kitty Hawk, Wilbur and Orville Wright could accurately be called aeronautical engineers. They were familiar with the science of the day and had not only added valuable information to that science, but also devised methods for exercising an unprecedented level of control over their craft during flight. They'd proven themselves capable of mastering the challenges that arose as they modified their plane and its design. The thrill and satisfaction of that day for them are difficult to imagine.

Rapid advances in flight technology would take place in Europe immediately after that first flight in 1903, but the Wright brothers' legacy was not just the successful flying of a powered, piloted heavier than air craft. It was the successful application of scientific principles to the testing of components for the purpose of controlled flight.

Notes

1. Anderson, J. D. (2002). *The airplane: a history of its technology.* Reston, VA, American Institute of Aeronautics and Astronautics.
2. Wright, W. a. O. (1909). Story of Our Lives. *New York Herald.* New York, New York.
3. (1903). Annual Report of the Board of Regents of the Smithsonian Institution Showing The Operations, Expenditures, and Condition of the Institution for the Year Ending June 30, 1902. *Doc. No. 484, Part 1.*
4. Wright, W., O. Wright, et al. (2001). *The papers of Wilbur and Orville Wright: including the Chanute-Wright letters and other papers of Octave Chanute.* New York, McGraw-Hill.
5. Congress, L. o. (1920). The Wilbur and Orville Wright Papers. *Legal Cases--Montgomery v. Wright-Martin Ariway Corp.-Depositions: Wright, Orville, 1920*, Manuscript Division.
6. Ibid.

7. Anderson, J. D. (2002). *The airplane: a history of its technology.* Reston, VA, American Institute of Aeronautics and Astronautics.
8. Ibid.
9. Wright, W., O. Wright, et al. (2001). *The papers of Wilbur and Orville Wright: including the Chanute-Wright letters and other papers of Octave Chanute.* New York, McGraw-Hill.
10. (1903). Annual Report of the Board of Regents of the Smithsonian Institution Showing The Operations, Expenditures, and Condition of the Institution for the Year Ending June 30, 1902. **Doc. No. 484, Part 1.**
11. Ibid.
12. Congress, L. o. (1900). The Wilbur and Orville Wright Papers. *Octave Chanute Papers, 1902–10,* Manuscript Division.
13. Ibid.
14. Ibid.
15. Ibid.
16. (1903). Annual Report of the Board of Regents of the Smithsonian Institution Showing The Operations, Expenditures, and Condition of the Institution for the Year Ending June 30, 1902. *Doc. No. 484, Part 1.*
17. Crouch, T. D. (1989). *The Bishop's boys: a life of Wilbur and Orville Wright.* New York, W.W. Norton.
18. Wright, W., O. Wright, et al. (2001). *The papers of Wilbur and Orville Wright: including the Chanute-Wright letters and other papers of Octave Chanute.* New York, McGraw-Hill.
19. Congress, L. o. (1900). The Wilbur and Orville Wright Papers. *Octave Chanute Papers, 1902–10,* Manuscript Division.
20. Ibid.
21. Wright, W., O. Wright, et al. (2001). *The papers of Wilbur and Orville Wright: including the Chanute-Wright letters and other papers of Octave Chanute.* New York, McGraw-Hill.
22. Anderson, J. D. (2002). *The airplane: a history of its technology.* Reston, VA, American Institute of Aeronautics and Astronautics.
23. Congress, L. o. (1912). The Wilbur and Orville Wright Papers. *Wind Tunnel--Correspondence, 1912–1928, 1938–1946,* Library of Congress, Manuscript Division, Washington, DC 20540.
24. Baals, D. D. and W. R. Corliss (1981). *Wind tunnels of NASA.* Washington, D.C., Scientific and Technical Information Branch for sale by the Supt. of Docs., U.S. G.P.O.
25. Jakab, P. L. (1990). *Visions of a flying machine: the Wright brothers and the process of invention.* Washington, Smithsonian Institution Press.
26. Ibid.
27. Ibid.
28. Ibid.
29. Ibid.
30. Administration, N. A. a. S. (2010). "The Beginner's Guide to Aeronautics." from http://www.grc.nasa.gov/WWW/k-12/airplane/bga.html.
31. Anderson, J. D. (2002). *The airplane: a history of its technology.* Reston, VA, American Institute of Aeronautics and Astronautics.
32. Parnell, G. S., P. J. Driscoll, et al. (2008). *Decision making in systems engineering and management.* Hoboken, N.J., Wiley-Interscience.
33. Jakab, P. L. (1990). *Visions of a flying machine: the Wright brothers and the process of invention.* Washington, Smithsonian Institution Press.
34. Baals, D. D. and W. R. Corliss (1981). *Wind tunnels of NASA.* Washington, D.C., Scientific and Technical Information Branch for sale by the Supt. of Docs., U.S. G.P.O.
35. Jakab, P. L. (1990). *Visions of a flying machine: the Wright brothers and the process of invention.* Washington, Smithsonian Institution Press.
36. Ibid.
37. Anderson, J. D. (2002). *The airplane: a history of its technology.* Reston, VA, American Institute of Aeronautics and Astronautics.

38. Congress, L. o. (1900). The Wilbur and Orville Wright Papers. *Octave Chanute Papers, 1902–10*, Manuscript Division.
39. Ibid.
40. Ibid.
41. "Centenniel of Flight."

Chapter 10
Rocketmen

*In spite of the opinions of certain narrow-minded people, who
would shut up the human race upon this globe, as within some
magic circle which it must never outstep, we shall one day
travel to the moon, the planets, and the stars, with the same
facility, rapidity, and certainty as we now make the voyage from
Liverpool to New York.*

Jules Verne, From the Earth to the Moon, 1865

The development of rocket technology is a story of international accomplishment.
It is also the story of three men working independently in three different countries who were at the forefront of liquid-fueled rocket development. All three
wrote extensively on the theory and design of rockets and their potential. All
three used models for testing of some kind. Two developed and tested actual
rockets. Due to political constraints, the work of each was not known to the
others, yet we know from their writings that all arrived at similar conclusions at
the start of the twentieth century.

Konstantin Tsiolkovsky

Konstantin Eduardovich Tsiolkovsky was born in Russia in the fall of 1857 in a
village south of Moscow. His birthplace was the fourth largest settlement in the
province at the time. The fifth of 18 children, it was his mother who taught him to
read and write. Tsiolkovsky first imagined a place without gravity at the age of 8
when his mother gave him a small hydrogen-filled balloon. He was intrigued by the
way it rose effortlessly to the ceiling each time he let it loose.

Tsiolkovsky lost most of his hearing due to scarlet fever and later wrote, "Age of
10 or 11, the beginning of winter, I rode a toboggan. Caught a cold. Fell ill, was
delirious. They thought I'd die but I got better, but became very deaf and deafness

G. Hagler, *Modeling Ships and Space Craft: The Science and Art of Mastering
the Oceans and Sky*, DOI 10.1007/978-1-4614-4596-8_10,
© Springer Science+Business Media, LLC 2013

wouldn't go. It tormented me very much." The hearing loss caused Tsiolkovsky to focus on individual pursuits, including his own education from the age of 14.[1] Even as an adult, he would describe himself as having set great goals for himself as a way to prove to himself and to others that he could excel.

Tsiolkovsky was 13 when his mother died. At the age of 16, Tsiolkovsky moved to Moscow to continue his studies through the use of the books at the Chertkovskaya Library. His father sent him money to live on, but it wasn't much. He studied mathematics, analytical mechanics, astronomy, physics, chemistry, and classical literature. "I ate just black bread, didn't have even potatoes and tea," he later remembered. "Instead I was buying books, pipes, sulfuric acid (for experiments), and so on. I was happy with my ideas, and black bread didn't upset me at all."

While pursuing his studies, Tsiolkovsky found a mentor in Nikolai Fredorovitch Federov. Federov was a Russian philosopher with some unique beliefs. He believed that men would eventually be forced to move into space because they would one day learn to bring the dead back to life. The resulting increase in the population would necessitate a move to other planets. Greatly influenced by Federov and wanted not only to go into space; he wanted to find a way for mankind to live and thrive there as they traveled freely among the planets.

"Around this time, Tsiolkovsky also discovered the novels of French science fiction and adventure writer Jules Verne. One novel that influenced him in 1865 was 'From the Earth to the Moon.' The impact on Tsiolkovsky was not unusual. It was with that novel that Verne first inspired a whole generation of spaceflight pioneers. 'I do not remember how it got into my head to make first calculations related to rockets,' Tsiolkovsky later wrote, 'It seems to me the first seeds were planted by famous fantaseour, J. Verne.' Unlike most of his contemporaries, however, Tsiolkovsky did more than simply marvel at Verne's descriptions of fantastic journeys. He questioned their practicality. He understood that shooting spacecraft from a giant cannon, Verne's method of reaching the moon, would inevitably kill its passengers due to the force of acceleration."[2]

In 1876, Tsiolkovsky found work as an apprentice teacher of mathematics, physics, and chemistry. He passed his exams and qualified as a schoolteacher in 1879. While in his twenties he began his research into life in space, balloons, and aerodynamics. By the time he was 21, he had independently articulated the basic principles of the kinetic theories of gases. He submitted his work to the Society of Physics and Chemistry in St. Petersburg. Although his theories were correct, they were nothing new. Tsiolkovsky worked in relative isolation. He had no way of knowing that this important work had already been done. However, the members of the Society, including Dmitry Mendeleyev, author of the Periodic Table, were duly impressed. Tsiolkovsky's second paper for the Society, *The Mechanics of a Living Organism*, earned him a place among its members.

In 1883, Tsiolkovsky published "Free Space." This work explored the possibility of living in outer space, along with the effects of zero gravity. He included a drawing of a spacecraft that could orient itself in space with the help of reactive jets rather than propulsive rockets.[3] This brought him to experiment with the possibility of reactive force. He used a container filled with compressed gas rather than solid fuel as his test case. The purpose of the experiment was to prove that an object could be propelled through the air by the power of escaping gas. To demonstrate this, he varied the movement of the container by varying the pressure of the gas released. His work proved to him that Newton's Third Law, which states that for every action there is an equal and opposite reaction, would apply to objects propelled through the atmosphere. This was significant because many doubted an object could propel itself by pushing against a gas. They were convinced and object required something solid to push against.

In 1890, Tsiolkovsky's paper, *The Problem of Flying by Means of Wings*, was the first to document his interest in aviation. By 1896, Tsiolkovsky drafted the design for a rocket that would be fueled by a mixture of liquid oxygen and liquid hydrogen. With his design, when the liquids were combined, they would create an explosion at the narrow, top portion of a chamber. The burning fuels would produce heated, condensed gas. The gas would be cooled, and then released through the tail of the rocket via a nozzle that would focus the stream of escaping gas and produce forward motion at a high velocity.

Also in 1898, Tsiolkovsky's *Elementary Studies of the Airship and Its Structure* was published. It was at this time that Tsiolkovsky built a wind tunnel to test the aerodynamic properties of a variety of aircraft designs for this work. With his wind tunnel, he studied the effects of friction and surface area on the velocity of the air over a streamlined shape. As a result of this work, the Academy of Sciences gave him funds, which he used to build a larger wind tunnel.

That same year Tsiolkovsky completed his most important work. It was an article, "Exploration of Space with Rocket Devices" that would be published in *Scientific Review* in 1903. In his article, Tsiolkovsky proposed the use of liquid-fueled rockets for the travel into space. He proposed liquid fuel because he calculated that it would provide more power than traditional solid rockets powered by gunpowder. The calculations he included showed that it would be possible to carry humans into space with "step rockets," known today as staged rockets. With such rockets, the lower rocket stage drops away after the fuel is spent, reducing the weight as the remaining sections of the rocket move progressively higher and attain escape velocity which allows the craft to fly beyond earth's atmosphere. Tsiolkovsky's stage design would permit far larger payloads to be carried aloft than would a single stage rocket powered by solid fuel.

Tsiolkovsky's calculation for reaching space by rocket is known as the "Tsiolkovsky Equation." It is a mathematical equation that relates the change in the speed of a rocket if no external forces act upon it with the effective exhaust velocity and the initial and final mass of the rock or other reaction engine. This equation was the first theoretical proof of the possibility of spaceflight. He would develop his ideas on rocketry and space travel further over the next 30 years. That work would include papers, monographs, and a science fiction novel of his own.[4]

Because of political unrest, the remote nature of the places he resided, and a lack of necessary materials, Tsiolkovsky was unable to build or test his rocket designs. Throughout his life, he did continue to write about space and what travel in space would mean. His writings include mention of the first space stations.

His notebooks contain drawings of humans in spacesuits passing through airlocks on a ship. He also wrote about the effects of weightlessness in space and envisioned a day in aviation when jets would be used to power aircraft. Since much of his work was self-published and few copies of the issue of *Scientific Review* in which his work appeared were available because the government confiscated most of them for an offense in another article, Tsiolkovsky's work was not well known in his country and certainly unknown outside his country. It wasn't until the international community recognized Goddard's work in the 1920s that the Soviet government published Tsiolkovsky's writings.

In 1926, Tsiolkovsky described 16 Stages of Space Exploration. These well thought-out, incremental steps were designed to bring man into space and allow him to thrive:

1. Design of rocket-propelled airplanes with wings.
2. Progressively increasing the speeds and altitudes reached with these airplanes.
3. Designing of a pure rocket without wings.
4. Developing the ability to land on the ocean surface by rocket.
5. Reaching of escape velocity and first flight into space.
6. Lengthening of the rocket flight time into space.
7. Experimental use of plants to make an artificial atmosphere in spacecraft.
8. Using of pressurized space suits for activity outside spacecraft.
9. Making of orbital greenhouses for plants.
10. Building of the large orbital habitats around the earth.
11. Using solar radiation to grow food, to heat space quarters, and for transport needs throughout the solar system.
12. Colonization of the asteroid belt.
13. Colonization of the entire solar system and beyond.
14. Achievement of individual and social perfection.
15. Overcrowding of the solar system and galaxy colonization.
16. The sun begins to die and the people remaining in the solar system's population move to other solar systems.[5]

Tsiolkovsky's vision for interplanetary travel included the use of gyroscopes to control the orientation of rockets in space; special pressure suits, airlocks, and tethers for those working outside their spacecraft; and a long-standing space station with many "Space Islands" that would sure as habitats for thousands of people. In 1929, Tsiolkovsky concluded that the first space flights would take place within 20 to 30 years.[6]

In the early 1930s Tsiolkovsky wrote several essays that were not published. These essays were about the existence of life throughout the universe. Tsiolkovsky believed this was possible because the atom was at the root of all existence and could take many forms. In keeping with the vision he'd formed as a teen in Moscow, he is reported to have told a friend in 1930, "For me, a rocket is only a means—only a method of reaching the depths of space—and not an end in itself... There's no doubt that it's very important to have rocket ships since they will help mankind to settle elsewhere in the universe. But what I'm working for is this resettling... The whole idea is to move away from the Earth to settlements in space."[7]

When Tsiolkovsky died in 1935, he left behind a body of work devoted to the attainment of his goal. His writings would be instrumental in the work of those who followed in the field of astronautics. He work would influence German rocket scientist Wernher von Braun, Russian rocket-engine designer Valentin Glushnko, and rocket designer Sergey Korolyov. Asteroid 1590, Tsiolkovskaja, is named after Tsiolkovsky's wife. The most prominent crater on the far side of the Moon is named for him. Tsiolkovsky is considered by many to be the father of astronautics. Interestingly, Tsiolkovsky considered himself a "Citizen of the Universe."[8]

Robert Goddard

American rocketeer Robert Goddard was born in Worcester, Massachusetts in 1882. He was inquisitive by nature and spent his childhood experimenting with a microscope and a telescope. He had a subscription to *Scientific American* and, as a teen, enjoyed the work of both Jules Verne and H. G. Wells. When he was 16 he read Wells' *The War of the Worlds*.

Two weeks after his 17th birthday, Goddard had what he would describe as a life-altering experience. He climbed a cherry tree behind the barn of his family home on October 19, 1899, the day he would forever after refer to as Anniversary Day, and later wrote, "I imagined how wonderful it would be to make some device which had even the possibility of ascending to Mars and how it would look on a small scale, if sent up from the meadow at my feet... It seemed to me then that a weight whirling around a horizontal shaft, moving more rapidly above than below, could furnish lift by virtue of the greater centrifugal force at the top of the path. In any event, I was a different boy when I descended the tree from when I ascended, for existence at last seemed very purposive."[9]

Goddard was a voracious reader, perhaps because he was frequently sick and confined to his home. His reading went beyond Verne and Wells, including Samuel Langley's scientific papers and Newton's *Principia Mathematica*. His study of Langley's work led him to focus his attention to birds in flight. Goddard's conclusion was that birds controlled their flight with their tails, not their wings as Langley suggested. His study of Newton's work led him to test the Third Law (for every action there is an equal and opposite reaction) to determine if it would apply to motion in the vacuum of space. His investigation satisfied him that it would, but it also led him to realize he needed a deeper understanding of mathematics and physics.

As early as 1907 "he prepared and submitted for publication a manuscript suggesting that heat from radioactive materials could be used to expel substances at high velocity from an orifice, thus furnishing jet propulsion sufficient to navigate in interplanetary space."[10] The article was not accepted for publication but "the idea of using hydrogen and oxygen as fuels for an interplanetary rocket, and the construction of such a rocket according to the multiple or step-rocket principle, occurred to him in 1909. After considerable calculation he put the theory into satisfactory form in the winter of 1912–1913. His computations included the possible use of smokeless powder for the propellant, as well as hydrogen and oxygen."[11] "Goddard was unaware of his Russian competition [Tsiolkovsky], and since he could not know that he had been trumped, he soldiered on."[12]

The first well-defined period of Goddard's rocket work extended from 1899 until after World War I. "It was a time of speculation, mathematical and theoretical development, and experiment with solid-fuel propellants."[13] By the time Goddard was a research fellow at Princeton in 1913, he'd come up with a formula to calculate the position and velocity of a rocket in vertical flight. His equation used the weight of the rocket, the weight of the propellant, and the velocity of the exhaust gases as variables. It was an important moment for Goddard but early that year he was forced to return home when he was diagnosed with tuberculosis. Originally thought to have only 2 weeks to live, he went on to live for another 30 years.

Perhaps his recovery and sustained good health can be attributed to his reliance on his own methods for restoring his health, "I shall never forget the smile and twinkle from the depths of a hammock, when there were several inches of snow on the ground and all the frost air was around zero. The thing that affected me now was that so far as we know he had no medical authority for his action (living on the veranda). It seemed to be his own idea entirely and he had in no uncertain terms absolved his family and his doctor from any responsibility." He improved steadily and by late March was back at work on his rocket theories.[14]

Goddard was quick to apply for patents to protect his intellectual property. He applied for two with the help of his father, while recovering from tuberculosis. U.S. Patent 1,102,653, issued in 1914, was for a step or multistage rocket. The patent covered "Rocket Apparatus" which was essentially everything to do with a rocket that was more complicated than a basic firework rocket or single stage projectile. The other patent he applied for was for a liquid fueled rocket. U.S. Patent 1,103,503 states that the fuel would be gasoline and liquid nitrous oxide.

By the fall of 1914 Goddard was well enough to accept a part-time position at Clark University in Worcester, Massachusetts. While there, he would be an instructor and research fellow. He used his research time to ready for his first powder rocket test in 1915. The tests were loud and distracting enough that Goddard was compelled to conduct future tests inside the physics lab. These tests proved that powder rockets were not very efficient, although their efficiency could be improved with the use of a nozzle. Still, it would not be enough to propel a significant payload to escape velocity and beyond.

That same year, Goddard devised an experiment to prove that a rocket could perform in the vacuum of space. Many doubted it would be possible that Newton's Third Law, every action has an equal and opposite reaction, would apply in a vacuum. They believed the rocket exhaust would be "sucked" out of the rocket and there would be no reactive force for propulsion. Goddard was determined to demonstrate that it would work. To prove this, he made a number of vacuum chamber experiments that set the question to rest for good when the rocket not only moved, but also operated at greater efficiency.

"More important... were experiments during 1915 that proved that a rocket would provide thrust in a vacuum. In the simplest form, Goddard set up a 0.22-caliber revolver loaded with blanks, mounted on an arm swiveling around a spindle, all in a bell jar from which the air had been withdrawn. When the pistol was fired by pulling a string attached to the trigger, the whole assembly twirled around. Just when he first devised this simple demonstration is not now apparent, but he repeated it often during the 1920s.

Science required rigorous measurements and impeccable experimental design to prove the point, so Goddard devised a large vacuum chamber with a rocket (at this stage, he usually called it a "gun") mounted inside, and fired when the chamber had been evacuated. Gauges measured the thrust caused by the rocket's lifting of the top of the chamber. Next, to prove that the motion observed was not the result of exhaust rebound, the experiment was repeated in a large circular vacuum tube.

More than 50 tests revealed that rockets provided about 20 percent more thrust in a vacuum than in air. More important, in Goddard's words, the work proved "that the phenomenon is really a jet of gas having an extremely high velocity, and is not merely an effect of reaction against the air"."[15]

Goddard again had no idea that Tsiolkovsky was had done similar work. Goddard was simply following the lead of his own intellect, in the same manner as Tsiolkovsky, as he worked to solve the problems of getting man into and moving man through space.

By 1916, Goddard needed additional funding to conduct his work. He approached the Smithsonian and, after providing them with a detailed manuscript of his work, received a $5,000 grant in January 1917. Clark University gave him a grant of $3,500 and Worcester Polytechnic Institute let him use their abandoned Magnetics Laboratory for testing. With this funding, Goddard entered the second well-defined period of his work during which he laid the experimental basis for his subsequent work and demonstrated the feasibility of liquid propellants for rockets.[16]

Unwanted Attention

The Smithsonian published the manuscript Goddard supplied in support of his grant application, *A Method of Reaching Extreme Altitudes*, in 1919. This book is a seminal work in the field of rocketry and was distributed worldwide. It laid out all of Goddard's mathematical theories of rocket flight, the work he'd done with solid-fuel rockets, and the possibility that man would someday be able to explore outer space. It also included a small section with Goddard's thoughts about a possible way to someday prove a rocket had actually reached the moon.

Unfortunately, the thought of traveling beyond the earth's atmosphere seemed outlandish to the general public in 1919. Up until then, Goddard had kept his ambitions in this realm largely to himself. His detailed thoughts about and calculations for a bright flash that would be visible from earth when the rocket reached the moon and crashed on its dark side, was simply too much for the press to ignore. They picked on this detail, not at all the focus of his manuscript, making it seem as if the entire work was about this one point. Any and every one with an opinion weighed in, including an editorial in the *New York Times* and an article in *Times* that called Goddard's understanding of Newton's Third Law into question. The *Times* article was wrong; it was Goddard who was correct and had proven it several years prior with his 1914 and 1915 experiments of his own design. That didn't keep the *Times* from writing that Goddard "only seems to lack the knowledge ladled out daily in high schools."

Goddard and Oberth

Goddard was understandably guarded in his communication with the press from that point on. He also was guarded when granting requests from scientists from other countries, possibly because he already saw the potential use of his work in weaponry. "Nils Riffolt remembered him routinely saying in the 1920s. "Let's keep this under our hat." He remembered also that the professor was free with information among his crew, and only somewhat less free with the press". Regarding a certain other party, Goddard was on guard: "Oberth was working in Germany, and Goddard didn't want to disclose too much. Goddard was outraged at Oberth's claim to independent invention, asserting that the German's work plagiarized his own 1919 Smithsonian paper."[17]

By 1921, Goddard was ready to test his liquid-fueled rockets. Powered with a gasoline and liquid oxygen mixture, his first test was on the grounds of his Aunt Esther's farm in Amherst, Massachusetts. His first successful launch of a liquid-propellant rocket took place on March 16, 1926. He was the first to ever use a liquid propellant for flight.

"In 1923 Goddard prepared a long, detailed report on all his tests, and an additional report asserting his priority, dating from 1899. He refuted Oberth's claims and alleged statements that Goddard's system could not reach into space. 'I do not wish to open the question of priority and thrash out all the phases of the matter in public,' he told the Smithsonian. Then he proceeded to do just that, point out flaws in Oberth's scheme for space travel, and evincing resentment of competition for first place in rocketry that, in the case of Oberth, bordered on paranoia."

Goddard addressed the annual meeting of the AAAS at the end of 1923. There he reviewed his work since 1909, refuting Oberth's contention that Oberth had been the first to consider hydrogen/oxygen propulsion in 1912. Goddard pointed out that only he held patents in rocketry, and reviewed the flaws in the German's "purely theoretical" approach, contrasted with his own actual tests.

He concluded: "This has been distinctly an American piece of work; it originated in America, as the writer's own interest and endeavors date back to 1899; the first theoretical work was published and the first experiments performed in America; and it seems very desirable that enough support be had to enable the work to be completed at an American laboratory."[18]

Goddard was clearly as aware of Oberth as he was unaware of Tsiolkovsky. As to the claims of all three, the best that can be said was that Tsiolkovsky was winding down just as Goddard and Oberth began.

Lucky Lindy to the Rescue

Goddard would continue tests of static and launched rockets, models of the rockets he hoped to someday test, until 1929 when the attention paid to each successive launch at his aunt's farm reached the point of it being a spectacle. Goddard clearly needed a more private place to conduct his research.

By 1929 Charles Lindbergh had decided the next logical step to aviation was likely to be rocketry. Lindbergh contacted Goddard a few months after one of his launches was covered in the press. The two met and hit it off, leading Goddard to be uncharacteristically open about his work. His trust in Lindbergh was well founded. Lindbergh was equally impressed with Goddard and ultimately convinced the Guggenheim family to fund Goddard's research. Their funding for the period from 1930 to 1934 would total $100,000, with more to come in the future.

This money enabled the Goddard's to move to Roswell, New Mexico. In Roswell, Goddard would be able to conduct his experiments in isolation and secrecy for over 10 years. During this third period of work, he would make significant progress in the development and flight of large gyro-controlled, pump-operated liquid-fuel rockets.[19] The Guggenheims supported Goddard's research, with one interruption, for many years. As part of his work in Roswell, Goddard ultimately experimented with gyroscopic guidance systems and parachute recovery.

Goddard considered even his failures a sort of success in that each taught him something about what would not work. In the tradition of Froude, Taylor, and the Wrights, Goddard's rockets grew progressively larger as the work done with smaller models served as the basis for the theory applied to larger models. By varying one or two variables at a time, he was able to build a store of data that informed the direction taken with each new iteration of his work.

Unfortunately, the U.S. Army never grasped the significance of Goddard's rockets for their purposes in his lifetime. It's also true that Goddard's rockets did not achieve significant altitudes but that was not the goal for Goddard. Goddard viewed each rocket prototype as a test case for the design of larger rockets. His goal was to perfect the efficiency and control of the smaller rockets before venturing into larger rockets. He was ready to build larger rockets, those capable of reaching "extreme altitudes," when World War II began.

Goddard died in August 1945, leaving behind groundbreaking and meticulous research that would inform the work of generations to come. "In his own thinking and that of his boosters, Goddard's achievements paraded as 'firsts,' and they made a formidable list. He first applied a De Laval nozzle to a rocket, and so redefined the rocket motor. He first proved that a rocket would work in a vacuum. He turned multi-stage and liquid-fuel rocketry into mechanical designs, not just ideas. He redefined battlefield rocketry, and provided the conceptual foundation for the bazooka. His 1919 publication on reaching 'extreme altitudes' was original, monumental, and elegant. He was the first inventor to launch a liquid-fuel rocket, the first to produce a rocket with an inertial guidance system, the first to use thrust-vector control (blast vanes) in a rocket, the fist to use a gimbaled engines, the first to build turbopumps for a rocket, the first to assemble liquid-fueled rocket motors in clusters, and the first to send a powered vehicle faster than the speed of sound."[20]

Because he was the first to launch liquid-fueled rockets, the first to control their flight, and the first to design powered vehicles that would break the sound barrier, he is considered by many to be the father of rocketry and space flight. Certainly, Goddard's work took the basic theories of fluid dynamics, developed and tested over centuries, into an entirely new realm. He used prototypes to achieve not only flight, but capable of escaping the atmosphere. His meticulous work was as seminal in the field of rocketry as the Wrights' work was in the field of aviation.

Hermann Oberth

Hermann Oberth was born in Romania, of German nationality, in 1894. His mother gave him a copy of Jules Verne's novel, *From Earth to the Moon*, when he was 11. He later recalled having read the book "at least five or six times and, finally, knew by heart." The calculations in the book led to his belief that space travel might be possible. By the time he was 15, Oberth's fascination with the book led him to envision a "recoil rocket." The rocket would propel itself into space by expelling the exhaust gases from a liquid fuel. This notion was shared by both Tsiolkovsky and Goddard, but Oberth had no way of knowing this at the time. He was not able to test his model but he continued to develop his theory by learning all he could of the mathematics he'd need when given an opportunity to turn his theories into reality.

Oberth realized that the higher the ratio between propellant and rocket mass, the faster his rocket would be able to travel. The obstacle to this higher ratio was the weight of the rocket itself. As the fuel was used, there was no need for the portions of the rocket that had carried and utilized that fuel, yet as long as it remained a part of the rocket, it would cause resistance and require a fuel output that was not justified for the utility gained. Unsurprisingly, Oberth's solution was to build a rocket with stages. As one section of fuel was exhausted, the physical portion of the rocket associated with that fuel would be ejected. As the mass of the rocket grew smaller, the rocket of the rocket would increase. Oberth wrote, "the requirements for stages developed out of these formulas. If there is a small rocket on top of a big one, and if the big one is jettisoned and the small one is ignited, then their speeds are added."[21]

After serving as a medic in the war, Oberth returned to the University to study physics. His doctoral dissertation on rocket-powered flight was rejected in 1922 because it was considered "utopian." Oberth later described his reaction in this way, "I refrained from writing another one, thinking to myself: Never mind, I will prove that I am able to become a greater scientist than some of you, even without the title of doctor. In the United States, I am often addressed as a doctor. I should like to point out, however, that I am not such and shall never think of becoming one."

Undaunted, Oberth used the dissertation as the basis for his book, *By Rocket into Planetary Space*. Highly technical and focused on the mathematical theory of rockets, the book also discussed the possibility of space stations and human inter-planetary travel. Oberth considered publishing a less technical version of his book but his workload did not permit it. Instead, a German spaceflight enthusiast, Max Valier, adapted his work for him. This book was well received by the public, and as a result rocket clubs were formed throughout Germany. The purpose of the clubs was to put Oberth's theories into practice. One club in particular, the VfR, attracted Oberth's attention during the 1920s. Among the members was Wernher von Braun, a man who would play an important role in the future of German and American rocket science.

Oberth's next significant work, *The Road to Space Travel*, was published in 1929. This 429-page book was authored by Oberth and dealt with ion propulsion and electric rockets. It was the first work to win the prize established by French rocket pioneer Robert Esnault-Pelterie at the French Astronomical Society. Oberth received the reward for his book and his encouragement of astronautics. He used his prize money to buy rocket motors for the VfR.

An unlikely booster of space travel at the time was the silent moviemaker Fritz Lang. Lang's movies were "must sees" in Germany and did much to promote the possibility of space travel to the intellectual elite. When Oberth was asked to consult on Lang's film, *The Woman in the Moon*, he readily agreed. Lang wanted everything to be realistic, right down to the spacecraft in the film. It's interesting to note that Lang was the first to use the modern day countdown. He used it when filming to keep everyone in synch for the simulated launch. When Lang decided a live launch would bring just the right publicity to his film, Oberth agreed to build a rocket for the purpose. Two days before the film premier, Oberth realized he could not com-plete the rocket in time and scrapped the project.

By 1938, Oberth was busy with research projects for Germany. He worked with von Braun on the V-2 development program during WWII. At the end of the War he settled in West Germany until 1955 when von Braun, then head of the U.S. army ballistic missile works in Huntsville, Alabama, invited Oberth to work for him. By 1959 Oberth had retired to West Germany perhaps because Oberth's vision for rock-ets and space was about more than their use in weaponry. In a 1954 article in *The American Weekly* magazine he said of unidentified flying objects (UFOs), "It is my thesis that flying saucers are real, and that they are space ships from another solar system. I think that they possibly are manned by intelligent observers who are mem-bers of a race that may have been investigating our earth for centuries…"[22]

In 1969, Oberth returned to the United States to witness the launch of the *Saturn V* rocket that carried the *Apollo 11* crew on the first lunar landing mission. Oberth died in December 1989, his influence having been felt in the rocket programs of both Germany and the United States.

A supporter of Oberth's wrote, "That Hermann Oberth is one of the three founding fathers of rocketry and modern astronautics is, I think, indisputable. That all three (Tsiolkovsky, Goddard, and Oberth) have advanced the science of rocketry is also indisputable—Professor Oberth, though, possessed a vision that set him apart, even from these great men." In 1923 he wrote in the final chapter of *Die Rakete zu den Planetenraumen (The Rocket into Planetary Space)*, "the rockets... can be built so powerfully that they could be capable of carrying a man aloft." In 1923, he became the first to prove that rockets could put a man into space. By all accounts Hermann Oberth was a humble man (especially considering his achievements) who had, in his own words, simple goals. He outlined them in the last paragraph of his 1957 book *Man into Space*: "To make available for life every place where life is possible. To make inhabitable all worlds as yet uninhabitable, and all life purposeful."[23] Because he was instrumental in the development of rockets in Germany, he is considered the godfather of early German rocketry.

It *Is* Rocket Science

Whether the rocket launched is large or small, its success depends upon the proper application of the forces of fluids in motion. The very same fluids in motion that apply to birds in flight, aquatic animals moving through water, vessels in the ocean, and airplanes in the sky. The science that led to the understanding and application of those principles to all of these phenomena is the same science that led to the theory and reality of rockets. Cayley's four forces of flight are also still in play (Fig. 10.1).

The science of Galileo and Newton are at the base of it all. Galileo conducted experiments involving motion. He also discovered the principle of *inertia*. This principle describes the way in which an object at rest resists changes to that state.

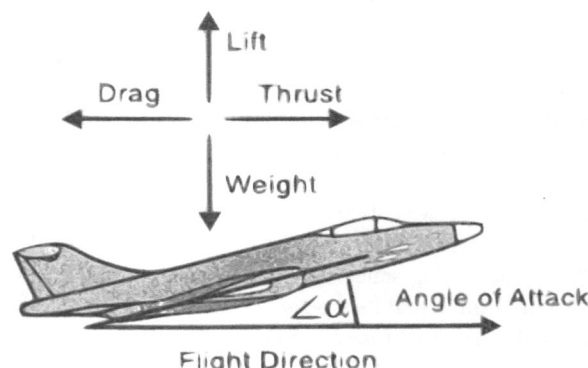

Fig. 10.1 Cayley's four forces of flight are Lift, Drag, Thrust, and Weight

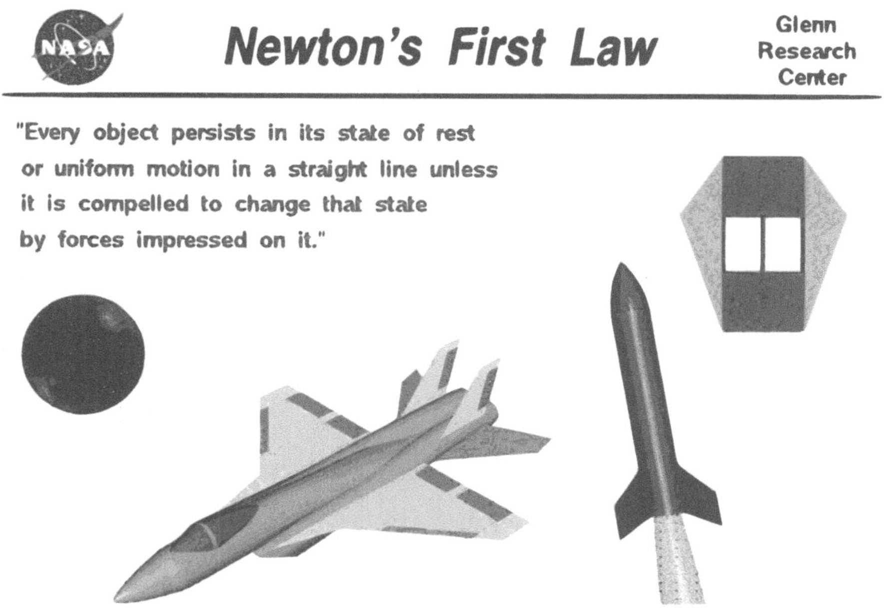

Newton's First Law

Glenn
Research
Center

"Every object persists in its state of rest
or uniform motion in a straight line unless
it is compelled to change that state
by forces impressed on it."

Fig. 10.2 Newton's First Law of Motion. *Source*: NASA/courtesy of nasaimages.org

The more the mass, the more the resistance. Those envisioning a rocket escaping the atmosphere must also envision a way to get the rocket up and running at the start.

Newton took Galileo's theories and the existing science of day one step further with the three laws of motion stated in his work, *Philosophia Naturalis Principia Mathematica*. The laws are straightforward: The First Law includes Galileo's Law of Inertia and states that objects at rest remain at rest and objects in motion remain in motion in a straight line unless acted upon by an unbalanced force (Fig. 10.2). The Second Law states that force is equal to mass times acceleration (Fig 10.3). The Third Law states that for every action there is an equal and opposite reaction (Fig. 10.4).

The concept of a *balanced force* is also important. Balanced forces are forces that are equal and opposite. Because they balance each other out, an object that is not moving is an object with a balanced force. When you stand on the sidewalk or sit on a chair, you are balanced by a force exerted on your mass that is equal to the force exerted by your mass.

Just as it's clear when a force is balanced, it's clear when a force is unbalanced because an object acted upon by an *unbalanced force* is an object remains in motion. It can be a rocket streaking for the sky or a coffee cup falling to the floor. The object is in motion because the force generated by its mass is not being opposed by an equal force. This can be a positive thing, as in the case of the rocket, because the force that is carrying the rocket skyward is generating *thrust* (Fig. 10.5). In fact, it

National Aeronautics and Space Administration

Newton's Second Law

Force = Change of Momentum with Change of Time

Difference form: $\quad F = \dfrac{m_1 V_1 - m_0 V_0}{t_1 - t_0}$ $\qquad$ t = time
$\qquad\qquad\qquad\qquad\qquad\qquad\qquad\qquad\qquad$ X = location

$\qquad\qquad\qquad\qquad\qquad\qquad\qquad\qquad\qquad$ m = mass

With constant mass: $\quad F = m \, \dfrac{V_1 - V_0}{t_1 - t_0}$ $\qquad$ V = Velocity

$$F = m \; a$$

Force = mass x acceleration

Velocity, acceleration, momentum and force are vector quantities

www.nasa.gov

Fig. 10.3 Newton's Second Law of Motion. *Source*: NASA/courtesy of nasaimages.org

  # Newton's Third Law   Glenn
Applied to Aerodynamics   Research
$\qquad\qquad\qquad\qquad\qquad\qquad\qquad\qquad\qquad\qquad\qquad\qquad\qquad$ Center

For every action, there is an equal and opposite re-action.

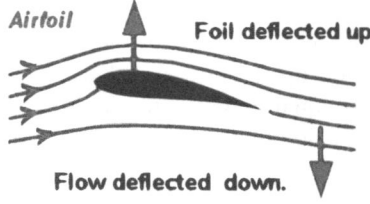

Airfoil   **Foil deflected up.**

Flow deflected down.

Ball deflected up.

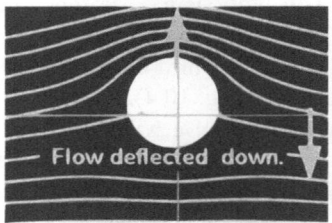

Flow deflected down.

Spinning Ball

Engine pushed forward.

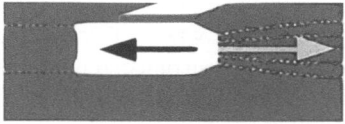

Flow pushed backward.

Jet Engine

Fig. 10.4 Newton's Third Law of Motion. *Source*: NASA/courtesy of nasaimages.org

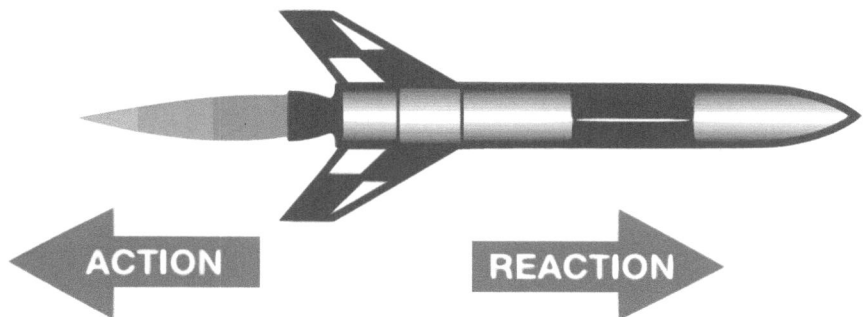

Fig. 10.5 Rocket thrust. Thrust is due to an equal and opposite reaction. *Source*: NASA/courtesy of nasaimages.org

is generating enough thrust to overcome the forces of gravity. This is an application of Newton's Third Law of Motion.

The amount of thrust a rocket generates will depend upon the factors described in Newton's Second Law. The force produced by the rocket is the thrust. It is directional proportional to the mass of the gas produced by burning the rocket propellant times the acceleration of the products of combustion out of the back of the rocket. Expressed as an equation: $f = ma$. The more propellant consumed (m) at any given moment, the greater the acceleration (a), the greater the thrust.

Once a rocket escapes Earth's gravity and makes it into space, the thrust that carries it there will be sufficient to keep that rocket in motion indefinitely, as explained by Galileo's Law of Inertia and Newton's First Law of Motion. Because there is no friction in space, hence no resistance or drag, there is nothing to slow the object in motion. Unlike every other area of fluid dynamics, in which minimizing resistance has been a primary consideration, a rocket that escapes Earth's gravity will never slow. It will not require additional thrust, either. The inertia it enjoys will carry it outward until it encounters an opposing force.

Not all forces acting upon a rocket will be the force that stops it. Forces can also be forces that alter the direction through the use of a *gimbaled nozzle*. The gimballed nozzle works with Newton's Third Law. It is a device that is as deceptively simple as a pitot tube. It acts to tilt the nozzle in different directions, causing the escaping force to exit at a desired angle from the present position (Fig. 10.6). Moveable fins are also used to alter the direction of a rocket in flight. These are another application of Newton's Third Law.

The largest obstacle to rocket flight and the manned travel through space it theoretically afforded was how to obtain the velocity required to clear Earth's atmosphere and how to direct the path of the rocket once it had. This is where the work with objects in a vacuum and ideas about solid and liquid propellants, and multistage rockets came into play. They were vital concerns if flight in space could ever be realized.

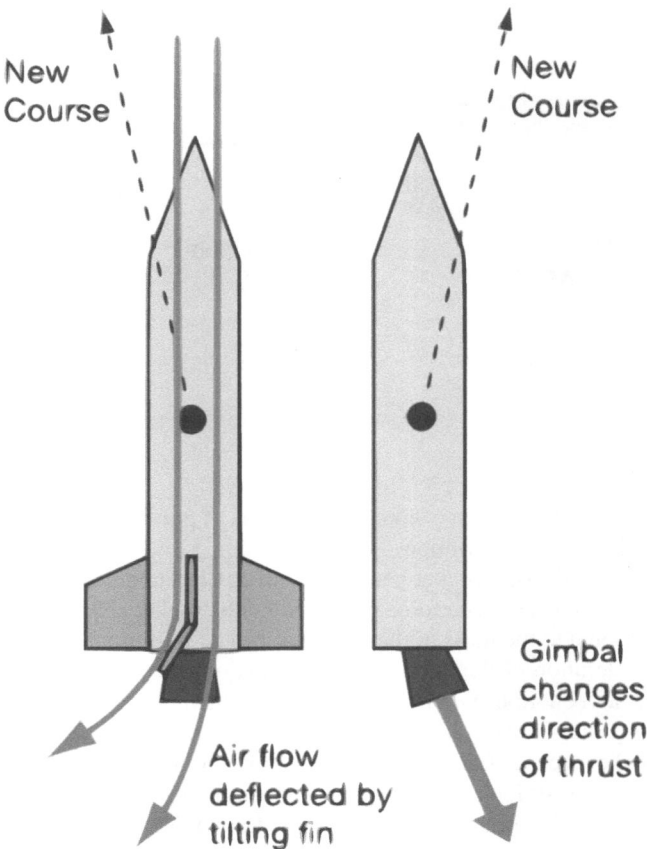

Fig. 10.6 The gimballed nozzle allows for a change in direction. *Source*: NASA/courtesy of nasaimages.org

The earliest rockets used solid propellants. When Wan Hu tried to launch himself it was with rocket stuffed with propellants. The top of the rocket was closed. The bottom is constricted by a narrow area known as the *throat* (Fig. 10.7). This constriction of the opening causes the escaping combustion products to accelerate greatly as they race outside. Solid core rockets must be ignited. They can be ignited through the use of a fuse—a dangerous proposition. They can also be ignited by igniting a source of flame that is packed inside the rocket, along with the solid propellant. This is the case with the space shuttle's SRBs.

Each of the Rocketmen (Tsiolkovsky, Goddard, and Oberth) theorized that liquid propellants could be used successfully with rockets. With liquids, a rocket will have a large tank within its body. One tank will contain a fuel like kerosene or liquid hydrogen. The other will contain liquid oxygen. When the liquid rocket engine is fired, they two are mixed as they are sprayed into the chamber (Fig. 10.8). This highly combustible mixture ignites, creating huge quantities of combustible

Fig. 10.7 A solid propellant
rocket. *Source*: NASA/
courtesy of nasaimages.org

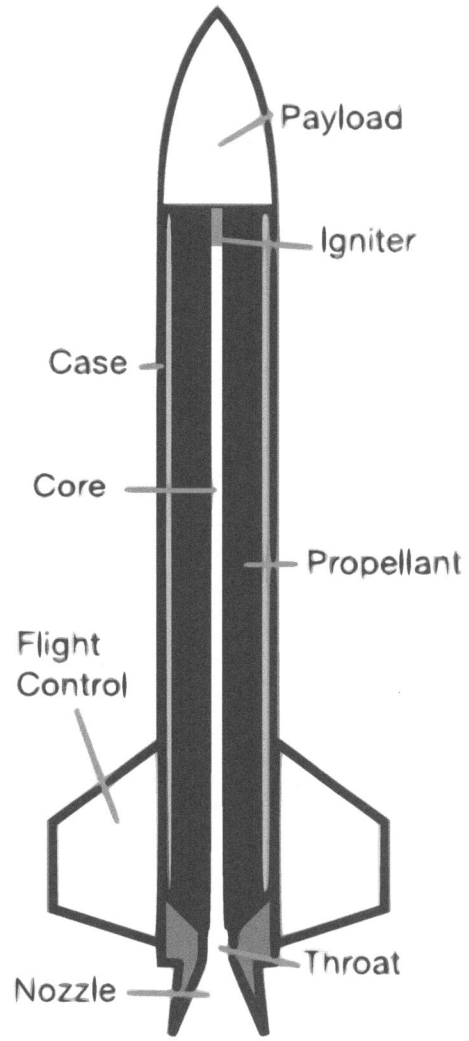

products. They shoot through the throat and are focused downward by the nozzle. Liquid propellant engines offer a degree of control that allows the thrust to be regulated by controlling the mixture and rate of spray.

All of the theory translates into reality once the rocket is a space vehicle with astronauts on board. Solid or liquid, the moment of lift off is a moment of anxiety. "So most astronauts getting ready to lift off are excited and very anxious and worried about that explosion—because if something goes wrong in the first seconds of launch, there's not very much you can do," said astronaut Sally Ride.

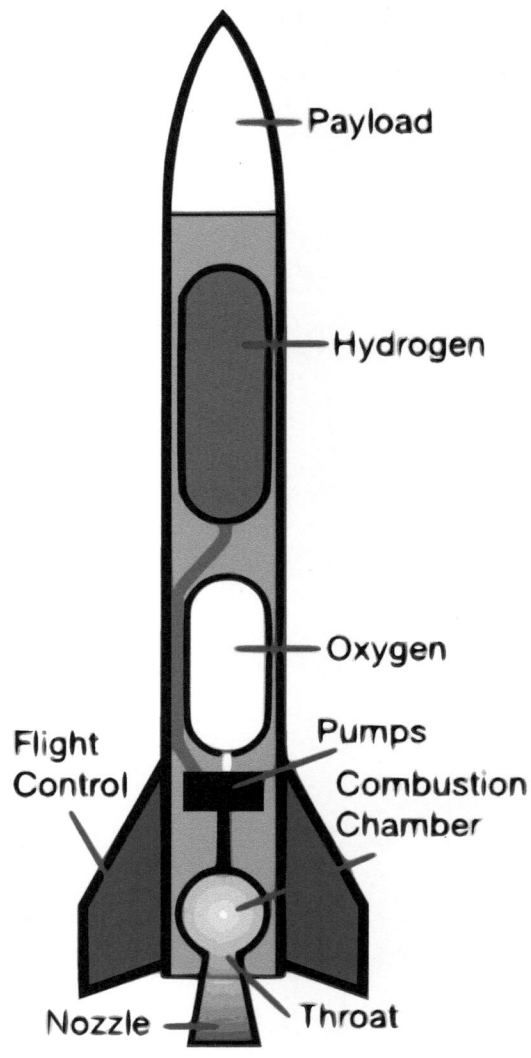

Fig. 10.8 A liquid propellant rocket. *Source*: NASA/ courtesy of nasaimages.org

With thrust and direction accounted for, that leaves only the resistance within the atmosphere as a consideration. Just as a whale must work harder while rising from the depths, a rocket must overcome the greatest amount of resistance at launch. A whale can inflate its lungs and use buoyancy to assist its upward motion. A rocket will benefit from the reduction in total weight as the propellant burns off, but it is still left with the physical structure of the rocket itself. This structure is subject to skin resistance in the same way as the wetted surface of a vessel is. If there were a way to minimize the total mass of the rocket as it traveled upward, it would allow the rocket to accelerate at a greater rate, resulting in greater speed and forward momentum.

The simplest answer to this was also anticipated by Tsiolkovsky, Goddard, and Oberth. Why not construct the rocket in such a way that a portion of the rocket could be made to fall away once it had completed its intended purpose? This is just the method in use today with the multistage rocket. As each component used in the generation of the thrust required to send the rocket soaring for its intended height is exhausted, it disengages and falls away. This reduces the total mass of the rocket and, in accordance with the theory of resistance, reduces the total resistance impeding the rockets forward motion.[24]

By balancing the four forces of flight, rockets blast skyward, subject to the same aerodynamic considerations of other creatures spending all or a part of their time moving through the air. They are certainly more elegant than most. They are absolutely moving at speeds that are unattainable for the rest. Yet their movement is subject to the same forces as something as insubstantial as an insect or a kite.

Once prototypes of full-scale rockets had been launched successfully, rocketry and space travel were no longer the province of lunatics. It wasn't long after that the idea of adding a payload containing humans became something to consider seriously. On May 25, 1961, travel to the moon became a national priority in the United States. In a speech on that day, then-President John F. Kennedy said, "First, I believe that this nation should commit itself to achieving the goal, before this decade is out, of landing a man on the moon and returning him safely to the Earth. No single space project in this period will be more impressive to mankind, or more important for the long-range exploration of space..."[25] It wasn't enough to send someone to space; they must come safely back. If this goal proved to be attainable, why couldn't the dream of interplanetary travel become a reality as well?

For those who first conceived of rockets heading into space, there had never been a question that travel to the moon was a worthy goal. After all, reaching the moon was the necessary first step on the path to far longer travels to far more distance planets and galaxies. Yet the question has been asked by those who are not drawn to the sky or have never felt the call of distant planets, *why head for the moon at all*? It's not enough to reply that it was the natural outcome of those first explorations into the air and beneath the sea. Perhaps a better answer is the one that was given by astronaut Neil Armstrong. "I think," he said, "we're going to the moon because it's the nature of the human being to face challenges. It's by the nature of his deep inner soul... we're required to do these things just as salmon swim upstream."

Conclusion

Tsiolkovsky, Goddard, and Oberth were instrumental in the development of rockets in their own countries. As their work became known by the international community, their theories influenced the work done in other countries as well. Their work was not done on full-sized rockets but on prototypes. They knew the areas they experimented with on their prototypes would have direct application to the larger rockets that would come after. The applicability was not in the strict sense that the use of scale models would ensure, but it was through the use of models that they explored the concepts essential to their broader theories.

One Goddard biographer sums up the influence of Tsiolkovsky, Goddard, and Oberth in this way, "the lives of the three "fathers" reveal some interesting similarities." All were inspired by Jules Verne. All were teachers oriented toward applied mathematics. All produced treatises demonstrating that projecting objects into space by means of liquid-fuel combustion was theoretically possible. Each had a streak of mysticism and indulged in fantasies about carrying the human species to the stars. Each had a flawed personality and difficulty getting along with others, and yet each was likeable enough to turn acquaintances into devoted admirers. Each had a monumental ego that approached the narcissistic, Tsiolkovsky less so than the others. All three lived to see the field each believed he had invented pass them by.

Such coincidences not withstanding, it is fair to assert that Goddard stands out among the three "fathers." If he had not made high-altitude rocketry both famous and respectable in 1920, the world might never have heard of either Tsiolkovsky or Oberth. Moreover, neither of the others ever launched a rocket, so only Goddard turned theory into practice. Fatherhood, it might be observed from biology, begins not with conception but at birth.[26]

Taking a broader view requires us to recall and recognize the centuries of unsung effort that went into the establishment of the theories behind the science of fluid dynamics. These theories were often based on the work of more than one man who might have been able to lay claim for the same discovery. For the Rocketmen who enjoyed the benefits of the work that had been done before their time, it can be said with authority that there are two components to their particular lasting legacy. One is that the power of an idea, in this case an idea eloquently brought to life in the form of a book by an author who wrote about things as he dreamed they might possibly be, ignited the imagination and passion of men who had never met. The other is that these men took their musings past simple thoughts of *what if* and moved them into the realm of the possible, changing the world in the process.

Notes

1. Sergeeva, G. a. T., Elena (2012). "Russian Space Web." 2012, from http://www.russianspaceweb.com.
2. Ibid.
3. Lytkin, V. "Konstantin Tsiolkovsky - The Pioneer of Space Travel." *Encyclopedia Astronautica.*
4. Ibid.
5. Ibid.
6. Ibid.
7. Kosmodemyansky, A. A. (1956). *Konstantin Tsiolkovsky: his life and work.* Moscow, Foreign Languages Pub. House.

8. Lytkin, V. "Konstantin Tsiolkovsky - The Pioneer of Space Travel." *Encyclopedia Astronautica*.

9. Clary, D. A. (2003). *Rocket man : Robert H. Goddard and the birth of the space age*. New York, Hyperion.

10. Goddard, R. H., E. C. K. Goddard, et al. (1948). *Rocket development; liquid-fuel rocket research, 1929–1941*. New York, Prentice-Hall.

11. Ibid.

12. Clary, D. A. (2003). *Rocket man: Robert H. Goddard and the birth of the space age*. New York, Hyperion.

13. Goddard, R. H., E. C. K. Goddard, et al. (1948). *Rocket development; liquid-fuel rocket research, 1929–1941*. New York, Prentice-Hall.

14. Clary, D. A. (2003). *Rocket man : Robert H. Goddard and the birth of the space age*. New York, Hyperion.

15. Ibid.

16. Goddard, R. H., E. C. K. Goddard, et al. (1948). *Rocket development; liquid-fuel rocket research, 1929–1941*. New York, Prentice-Hall.

17. Clary, D. A. (2003). *Rocket man: Robert H. Goddard and the birth of the space age*. New York, Hyperion.

18. Ibid.

19. Goddard, R. H., E. C. K. Goddard, et al. (1948). *Rocket development; liquid-fuel rocket research, 1929–1941*. New York, Prentice-Hall.

20. Clary, D. A. (2003). *Rocket man: Robert H. Goddard and the birth of the space age*. New York, Hyperion.

21. "Hermann-Oberth Raumfahrt -Museum." from http://www.kiosek.com/oberth/.

22. Walters, H. B. (1962). *Hermann Oberth: father of space travel*. New York, Macmillan.

23. "Hermann-Oberth Raumfahrt -Museum." from http://www.kiosek.com/oberth/.

24. NASA. "NASA." *Educator Guide - Rockets*. from http://www.nasa.gov/pdf/280754main_Rockets.Guide.pdf.

25. Benson, T. (2009). "NASA." *Glenn Research Center*. 2012, from http://www.grc.nasa.gov/.

26. Clary, D. A. (2003). *Rocket man: Robert H. Goddard and the birth of the space age*. New York, Hyperion.

Part IV
Model Testing Today

Chapter 11
Computational Fluid Dynamics

> *The physical aspects of any fluid flow are governed by the*
> *following three fundamental principles: (1) mass is conserved;*
> *(2) F = ma (Newton's second law); and (3) energy is conserved.*
> *These fundamental principles can be expressed in terms of*
> *mathematical equations, which in their most general form are*
> *usually partial differential equations. Computational fluid*
> *dynamics is, in part, the art of replacing the governing partial*
> *differential equations of fluid flow with numbers, and advancing*
> *these numbers in space and/or time to obtain a final numerical*
> *description of the complete flow field of interest.*

John D. Anderson, Computational Fluid Dynamics: An
Introduction

Innovators have long turned to technology for better ways to perform familiar tasks. For William Froude, that meant creating a controlled environment for his scale model experiments. David Taylor wanted a similar environment for his work, taking Froude's results as a starting point and pushing forward to new levels of accomplishment. Not long after, the Wilbur and Orville Wright realized they could use models in a controlled environment to perfect the design of their flying machines. When the time came to head for the moon, Robert Goddard was there with his rocket prototypes.

Each of these men built upon the theories, science, and discoveries of those who came before them, but innovation and new technologies did not end with them. New challenges arose with the increasing complexity of weapons like the intercontinental ballistic missile (ICBM) and vehicles for the peaceful manned exploration of space. These new vessels could not be readily tested. The calculations involved in predicting the performance of blunt bodies (those with a rounded nose) like the ICBM were too complex for the methods that had worked in the past. A new technology was needed, one that would allow the virtual modeling of behavior in a fluid field.

That new technology arrived in the form of the computer in the late 1950s and early 1960s. With the computer, it was possible for calculations to be done in a

G. Hagler, *Modeling Ships and Space Craft: The Science and Art of Mastering*
the Oceans and Sky, DOI 10.1007/978-1-4614-4596-8_11,
© Springer Science+Business Media, LLC 2013

fraction of the time they normally would have taken. But which calculations were the important ones and how could the results be judged?

Virtual Testing

Virtual testing makes use of existing data to arrive at predictions about the behavior of a design. The better the fit of the new design to the old design, the more reliable the outcomes will be since the models that are used to perform the calculations are based on the current body of relevant knowledge. That is why scale model tests are still made today. The biggest change is that these physical tests are generally made near the end of the design cycle. Especially in the case of wind tunnel testing where the cost of the testing can be prohibitive, using CFD for the preliminary analysis can clearly be the most efficient way to arrive at a new design.[1]

As a result of this, CFD "has become so strong that today it can be viewed as a new 'third dimension' in fluid dynamics, the other two dimensions being the classical cases of pure experiment and pure theory."[2]

Methodology

CFD models measure performance in one state. For example, they are designed to measure performance in water, not in situations where water is turning into steam. Because the models use a steady state, Navier–Stokes equations are used for most CFD applications. If viscosity is not a factor, that variable is removed and the Euler equation is used instead.

Two-dimensional models are used for the design and analysis of airfoils. These models also include boundary layer analysis and have been in use since the 1930s. As computer power increased, modeling software moved on to three-dimensional calculations. In 1967, the first paper written on a practical three-dimensional method to solve what are known as "linearized potential equations" involved the transfer of continuous models and equations to discrete models and equations. Because the method focused on the surface of the geometry with panels, these types of models are called Panel Methods. The first model did not include lifting flows and was applied to ship hulls and aircraft fuselages.

Since that time, a number of different types of models have been put into use. Because all the calculations must be done in real time, on numerous variables, the computing requirements are significant. As technology advances, some of the models can be run on relatively smaller computers. The most sophisticated models still require the use of super computers and the need for more powerful computers for these purposes places significant pressure upon those at the forefront of computer design to meet the need.

CFD at Universities

CFD is no longer the realm of a few facilities with access to the world's largest computers. Many smaller-scale versions of CFD models are available and are being used by students at universities to enhance their intuitive understanding of the behavior of fluid flows. Putting the ability to model a sophisticated interaction like that occurring in the boundary layer of an airfoil then opens the possibility of applying the same modeling capabilities to problems that are not aerodynamic in nature but still utilize the equations and principle associated with aerodynamics. It also opens the way for models utilizing equations specific to other fields of investigation. By putting CFD models in the hands of students and researcher alike, universities are expanding the base of professionals who are familiar with this technology and the reliable and meaningful data it provides.

What Is Modeled?

Just what is being modeled? Any type of fluid flow can be modeled, given enough computing power and an algorithm that contains the pertinent variables. The underlying equations of CFD, and of a particular CFD model, are known to the investigator. By utilizing a model with a basis that best meets the information necessary for the desired outcome, researchers can enhance the reliability of their findings. That's not the idea behind the models, though. These models are also used to keep the cost of testing down by reducing the time from concept to selection of the final design. They do this by making it possible to leave the actual physical testing of a scale model, a costly and time-consuming process, to the end of the process, when only the design or designs with the most potential are still under consideration.

It's also unnecessary to test known hull designs or fuselage designs each time. With CFD only the changes will be tested to help determine whether or not a new design will enhance performance. It's important to note that physical testing of scale models is still done at the end of the process for a radically new design as a way to verify the results obtained through the CFD process. Those results will then become part of the knowledge base that is applied to the next new design.

When designing vessels, CFD can be used to see the effect of a narrower overall width when coupled with a larger or smaller draft. It can help determine the optimal configuration of vessels that will operate in shallow water or identify the most important considerations for a vessel that must operate equally well in shallow and deep water.

CFD is used to simulate future flow over a vehicle. It allows the observer to define the flow at a specific grid point. It can solve the problem of flow over an airfoil, display laminar and turbulent pipe flows, aspects of a boundary layer, and steady and unsteady flows past a cylinder.

When designing aircraft, CFD can be used to simulate jets in operation at various Mach levels or during maneuvers that are too risky or not even possible with current technology. Innovative helicopter designs can be tried out in a virtual environment to determine which has the greatest potential for sustained investigation. Civilian aircraft designs can also be tested with CFD models to help determine the risk and economic trade offs inherent in new designs or larger planes. Simulation of the flow over an airfoil at various angles of attack is also possible, leading to the ability to test the stall angle of new wing designs.

Designs for spacecraft, from rovers to launchers, can also be modeled with CFD. One application is to estimate the forces on a vehicle during reentry. The new commercial craft being developed to take passengers up to zero gravity conditions before a return to earth can be modeled in this way, too. There are even reports that SpaceShip One was designed, built, and flown without any physical modeling or testing at all.

More than Ships, Planes, and Rockets

CFD is used to model more than just the craft that make their way through the sky. It's being used to model the boundary layer of submarines. It's finding a home for the visualization of the action of the circulatory system in the human body to determine the effect a new type of shunt might have. CFD can give insight into the impact a pacemaker has when implanted in the body. An investigation into fluid flows can help researchers study the effects microsurgery has on systems that include fluid flow. Systems like the human heart.

As the cost of CFD testing comes down with the advent of new software and increased computing power, new designs for automobiles and skyscrapers can also be tested with CFD models. The same is true for bridges. Each of these is subject to the effects of fluid flows. The more that can be learned about the effect of these flows in the design phase, the better the ultimate design will be.

Fire flow models are currently under development. These will help determine the likely path of wildfires. As more and more people choose to build homes in close proximity to areas prone to wildfires, these models will enable firefighting agencies to deploy their men and equipment at the most advantageous locales. They also may result in some different methods for fighting these massive fires.

Weather patterns are the result of massive fluid flows. The ability to predict the weather is not just important to the daily life of millions of people. It's also important for strategic and military reasons. Massive amounts of computing power are necessary for work with these flows.

The primary criterion for the use of CFD is whether or not what is being studied is subject to the principles of aerodynamics. Since aerodynamics is the science of a fluid in motion, it stands to reason that complex problems in associated areas will find the use of CFD models expedient. Giving scientists and other researchers a new method for interpreting the behaviors they observe will lead to better designs with greater control over and use of the principles of fluid dynamics.

Scale Models

Scale models still have a place in the high tech world of today. They begin as virtual models, being tested with algorithms in a virtual world. Each detail of the models can be made to mimic the physical models that would have been built by Froude, Taylor, the Wrights, or the first astronautic theorists.

The principles expounded and proven by Froude are the foundation of it all. Whether virtual or physical, necessary and accurate data can be obtained from the use of scale models. Physical scale model testing is especially important in the case of a brand new design. These physical scale models are tested in wind tunnels and model basins, just as they were at the start of the twentieth century, to not only test the design, but also to validate the findings of the CFD model. It's an iterative process and each informs the function of the other.

There is much value in the use of scale models to test conditions and performance, and not just in wind tunnels and model basins. Robotic vehicles and rovers are made to scale and moved through simulated environments to see how they function. These are not necessarily robots that will be tested in CFD models because they do not necessarily involve fluid flows, but their interaction with the fluids can be modeled. Whatever the results, when taken from the model and made to full scale, the findings will be trusted and valid because Froude's Law of Comparison has stood the test of time.

Notes

1. Wendt, J. F., J. D. Anderson, et al. (2010). *Computational fluid dynamics: an introduction.* Berlin; New York, Springer-Verlag.
2. Ibid.

Glossary

Acceleration A change in motion. Can be an increase in speed or a change in direction.

Action The result of a force.

Aerodrome The name Samuel Pierpont Langley used for his airplanes, from his earliest gliders to his final attempts at flight.

Aerodynamic forces The four forces that act on an object in flight: lift, gravity, drag, and thrust. First identified by Cayley in 1799. The basis for the configuration of modern, fixed-wing aircraft.

Ailerons Used to rotate the craft about the longitudinal axis. Mounted on the trailing edge of each wing near the wingtips and moving in opposite directions. Help to maintain lift.

Airfoil The cross-sectional shape of an airplane wing.

Alveoli Small air-containing compartments of the lungs.

Angle of attack (α) Angle at which relative wind meets an airfoil.

Archimedes' principle When an object is immersed in a fluid, it is buoyed upward by a force equal to the weight of the fluid displaced.

Aspect ratio (AR) The ration of the span to the chord of the wing. It is calculated as the square of the span divided by the wing area.

Balanced force Forces equal each other. An object is at rest.

Bernoulli's principle As the velocity of a fluid increases, the pressure exerted by that fluid decreases.

Blubber The fat of whales and other large marine mammals. It is less dense than seawater and is a source of buoyancy for marine mammals.

Bony fish Any of a major taxon comprising fishes with a bony rather than a cartilaginous skeleton.

Boundary layer The region between the wall of a flowing fluid where the fluid is not flowing at all and the point in which the flow speed is nearly equal to that of the fluid. The area where friction occurs and causes resistance.

Bow wave The wave that forms at the bow of a ship as it moves through the water. This can be mitigated by a bulbous bow.

G. Hagler, *Modeling Ships and Space Craft: The Science and Art of Mastering the Oceans and Sky*, DOI 10.1007/978-1-4614-4596-8,
© Springer Science+Business Media, LLC 2013

Bulbous bow A bulb that extends forward of the bow, beneath the surface of the water. Works to lessen *resistance*.

Buoyancy The tendency of a fluid to exert an upward force, equal to the weight of the fluid displace, on a body placed in it.

Camber A measure of the curvature of the airfoil. The higher the camber, the greater the curvature. Cambered airfoils are aerodynamically efficient.

Center of buoyancy The center of gravity of the immersed part of a ship or floating object.

Cetacean Whales, dolphins, porpoises, and related forms; among other attributes having a long tail that ends in two transverse flukes.

Chord The width of a wing.

Coefficient The number in front of a variable. Expresses the relationship to that variable.

Coefficient of viscosity The ratio of the shearing stress to the velocity gradient.

Compressibility The ability to become denser when pressure is applied.

Compressible A compressible substance will take up less space when pressed.

Continuity equation AV = constant where A = the area and V = velocity. When a fluid is in motion, it moves so that mass is conserved. If the area is reduced, the velocity increases and vice versa. da Vinci.

Continuously Flowing without a break or gap in the flow. An essential of a Newtonian fluid.

Continuum Molecules that are in close relation to one another without breaks or gaps.

Cruciform tail Looks like a cross when viewed from the rear.

Density A measure of how tightly packed a substance is. Considered constant for incompressible fluids.

Dermal denticles Dermal denticles are sharp, V-shaped scales on a shark's skin. They decrease drag and turbulence. Also called *placoid scales.*

Dihedral The upward angle of a fixed-wing aircraft's wings. Observable with the Turkey vulture when in a glide.

Dimensionless figure Having no units associated with it. The same under all possible systems of units. The Froude number is one example.

Displacement The property of a body immersed in a fluid to push the fluid out of the way and occupy the space.

Dorsal At or relatively near the back and corresponding surface of the head, neck, and tail.

Drafting A technique in which two moving objects align in a close group to reduce the overall effect of drag due to exploiting the lead object's slipstream. Used by Canada geese in v-formation and cyclists riding in a Peloton.

Drag One of the four aerodynamic forces. It is the amount of change in horizontal momentum caused by resistance of the medium around the object in motion. It is the force that is opposite *thrust*. Also known as *resistance*.

Drag coefficients (cd or cx or cw) A dimensionless quantity used to quantify the drag of an object in a moving fluid.

Elevator Mounted on the trailing edge of the horizontal stabilizer on each side of the fin in the tail. Move up and down together. Alters the angle of attack of an airplane.

Energetics A method to quantitatively assess the effort animals spend acquiring resources, as well as the relative way in which they allocate those resources.

Euler's equations A set of equations governing inviscid fluid flows—those that are without friction.

Fineness ratio (FR) Ratio of body length to maximum diameter. A value of 4.5 is considered to provide the least drag and surface area for the maximum volume.

Fins A surface used for stability and/or to produce lift.

Flipper Flattened forelimb of a marine mammal.

Fluids State of matter in which a substance cannot hold a shape. They flow continuously. Can be air, gas, or water.

Flukes Horizontally spread tail of a whale.

Force Push or pull exerted on an object

Force coefficients (*m*) A dimensionless force in a specified direction. Can be used for lift and drag calculations.

Four forces of flight Lift, gravity, drag, thrust. Sir George Cayley. By breaking them apart, a different method could be used for the generation of each force.

Froude number (*Fr*) A dimensionless number that describes the flow pattern over an obstacle.

Fusiform Having a spindle-like shape that is wide in the middle and tapers at both ends. A teardrop shape that is hydrodynamically and aerodynamically efficient.

Gimbal Pivoted support. Allows motion around a single axis. Can change the direction of thrust on a rocket.

Glide To move in a smooth, effortless manner.

Gravity One of the four aerodynamic forces. The downward force that is also known as weight. It is the force of attraction between two objects. It is the opposite of *lift.*

Helicoidal twist The sort of twist exhibited in DNA.

Hydrostatic pressure The pressure exerted by a fluid at equilibrium due to the force of gravity. This is the pressure exerted in a water column.

Ideal fluid A fluid without internal friction or viscosity. It does not exist in reality and is an imaginary fluid. (Inviscid fluid)

Incompressible An incompressible substance does not take up less space when pressed.

Inertia The resistance of a mass to changes in motion. The more the mass, the more the resistance. Galileo.

Internal friction A measure of the attraction between the molecules of a substance.

Inviscid fluids Fluids that have no internal friction or viscosity. These ideal fluids do not exist in reality.

Laminar fluid flow The streamlines in an ideal fluid.

Laminar regime An instance of layered flow without turbulence.

Lateral axis Passes through the aircraft from wingtip to wingtip. *Pitch* is rotation around this axis.

Lateral control Control from wingtip to wingtip.

Law of comparison $V = k(L)^{1/2}$ Scaling between a model and the actual vessel applies when the speed (*V*) is proportional to the square root of the length (*L*) and *k* is the number that applies to both model and prototype.

Law of continuity $AV=$ constant. Applies to incompressible fluids. da Vinci.

Lift It is the force that directly opposes the weight of an airplane and the force of gravity and holds the aircraft aloft. It is calculated as the pressure factor × velocity squared × wing area × lift factor.

Lift coefficient (cl) The ratio of the object's lift to the drag of a perpendicular flat plate with equal area.

Lilienthal's tables Otto Lilienthal created a table with the lift coefficients under varying conditions. These figures were based on his research. The Wrights used his figures and found that they were "wrong." Actually, they were correct for the wing shape used by Lilienthal, but their efforts to replicate Lilienthal's figures spurred the research that led to their successful flight.

Liquid propellant Propellants in liquid form.

Longitudinal axis Passes through the aircraft from nose to tail. *Roll* is rotation around this axis.

Longitudinal control Control from nose to tail.

Magnus force The lift force, first observed by Gustav Magnus in 1853, that acts on a spinning object in motion. A variant of the Bernoulli Principle.

Marine mammals A mammal that lives in a marine or aquatic environment.

Mass Amount of matter contained in an object. Does not have to be solid.

Metacenter The intersection of vertical lines through the center of buoyancy of a floating body when it is at equilibrium and when it is floating at an angle. The location of the metacenter is an indication of the stability of a floating body.

Model basin A facility where everything is built to scale for the testing of scale models.

Momentary equilibrium That which would place the axis of equilibrium normal to the wave surface at the point where it floats.

Navier–Stokes equations Equations that describe the relationship of velocity, pressure, temperature, and density of a moving fluid.

Newton's first law Every object persists in its state of rest or uniform motion in a straight line unless it is compelled to change that state by forces impressed on it.

Newton's second law The change in velocity with which an object moves is directly proportional to the magnitude of the force applied to the object and inversely proportional to the mass of the object.

Newton's third law For every action, there is an equal and opposite reaction. At one time incorrectly believed to be the sole source of lift.

Newtonian fluids These fluids quickly correct for shear stress when the object causing the stress is no longer present.

Non-Newtonian fluids Shear stress causes a fundamental change in the makeup of the liquid for some period of time after the stress is removed.

Observable surface The flat surface at the highest level of a fluid when it does not completely fill its container.

Ornithopter An aircraft that flies by the flapping of wings.

Parabolic foils Foils with inward curving sides.

Paraffin A type of wax that holds a shape but is easy to mold.

Parametric study A study in which one variable is changed to note the effect on an object.

Patagium The skin covering the surface of the wing of a bat. The fold of skin connecting the forelimbs and hind limbs of flying squirrels and some other tetrapods. The fold of skin in front of the main segments of a bird's wing.

Pinniped Seals and sea lions.

Pitch A change in the vertical direction of the aircraft's nose. Rotation around the lateral axis.

Pitch axis The lateral axis through the plane from wingtip to wingtip.

Pitot tube A simple device that measures flow velocity by measuring the difference between the static and dynamic pressures in a fluid. Used to determine the speed of an aircraft.

Placoid scales Sharp, V-shaped scales on a sharks skin. They decrease drag and turbulence. Also called *dermal denticles*.

Planar surface A two-dimensional surface that is perfectly straight in each dimension.

Planform The shape of a wing or other object when viewed from above.

Planophore The name Alphonse Penaud gave to his model airplane. It was the first airplane to exhibit longitudinal stability, due to the angle of the tail and the location of the wing.

Porpoising Leaping at least partly clear of the water surface during rapid swimming.

Principle of similitude The Law of Comparison when applied to the resistance of ships. An essential outgrowth of William Froude's work in the model basin at Torquay. The essential principle at work in the use of scale model testing.

Propellant Material that produces pressurized gas that can be directed through a nozzle to produce thrust.

Propulsion To push or drive an object forward.

Pusher propeller Propeller mounted behind the wing that push, rather than pull, the airplane through the air.

Reaction What happens when the action occurs. The recoil on a gun. Air rushing out of a popped balloon.

Real fluids Defined as having internal friction or viscosity. (Viscous fluid)

Residual volume (RV) The volume of air remaining in the lungs after a maximal exhalation.

Resistance One of the four aerodynamic forces. It is the amount of change in horizontal momentum caused by resistance of the medium around the object in motion. The force opposing *thrust* or forward motion. Also known as *drag*.

Reynolds number (*Re*) A dimensionless number that expresses the ratio of inertial forces to viscous forces.

Ridge lift Created when wind strikes an obstacle that deflects the wind upward.

Roll A change in the orientation of the aircraft's wings with respect to the force of gravity. Rotation around the longitudinal axis.

Roll axis The longitudinal axis that passes through the plane from nose to tail.

Rudder Primary means of controlling yaw. Controls the rotation around the vertical axis.

Schooling Fish swimming in the same direction in a group.

Shear stress A stress in a fluid that is parallel to the fluid motion velocity or streamline.

Slipstream A pocket of reduced pressure following an object moving through a fluid.

Smeaton's coefficient (k) The drag of one square foot plate at one mile per hour. (Mistakenly thought to be 0.005 in 1900.)

Solid propellant Propellant is a solid and packed into the rocket before lift off.

Soliton A solitary traveling wave.

Span of a wing Distance from wing tip to wing tip

Squalene A colorless unsaturated aliphatic hydrocarbon, $C30H50$, found especially in human sebum and in the liver oil of sharks.

Stearine A type of wax that holds a shape but is easy to mold.

Streamline A continuous series of particles that follow one another in an order fashion in parallel with other streamlines.

Streamline fluid flow At all points in the flow, the fluid flows are in the direction of the fluid velocity.

Streamline theory Fluids flow uniformly in the direction of the fluid velocity. Based on Bernoulli's work. Put forth as a theory by Rankine. Replaced the waveline theory.

Streamlining A fusiform design that reduces drag.

Taylor standard series Created by David Taylor. Comprised of 80 models with systematically varying proportions.

Thrust It is the force that moves an object through a fluid. It is the opposite of *drag* or *resistance*.

Tidal volume The amount of air breathed in or out during normal respiration.

Torque effect An equal and opposite reaction tending to rotate the object in the opposite direction.

Total lung capacity (TLC) The maximum volume to which the lungs can be expanded with the greatest possible inspiratory effort. It is equal to the vital capacity (VC) plus the residual volume (RV).

Tractor propeller Propellers mounted ahead of the wing and oriented to pull rather than push the airplane through the air.

Turbulent fluid flow A fluid flow in which the streamlines are disrupted.

Turbulent regime An instance of flow in which the layers are not in streamlines and shear stress is present.

Unbalanced force Forces are not equal. Object is in motion.

Velocity gradient Defined along a channel cross-section, it is the difference in velocity from the bank to the local peak velocity divided by the distance between those two points.

Velocity-squared law Resistance is not proportional to the velocity; it is the square of the velocity.

Vertical axis Passes through the aircraft from top to bottom. *Yaw* is the rotation around this axis.

Viscosity The measure of internal friction in a fluid.

Viscous Having internal friction.

Viscous fluids Fluids that have internal friction.

Viscous fluid flows Fluid flows with viscosity sufficiently large to make viscous forces a significant part of the total force field in that fluid.

Vital capacity (VC) The maximum tidal volume.

Vortex The rotating motion of a fluid around a central core.

Water column A theoretical column of water that extends from the surface to the bottom sediment.

Waveline system A vessel design theory that takes the length and shape of the wave into consideration.

Waveline theory Theory of the motion of a vessel through water. Proposed that the vessel literally pushed the water out of the way, essentially carving a channel in the water as it progressed. Scott Russell.

Weight One of the four aerodynamic forces. The downward force that is also known as *gravity*. The opposite of *lift*.

Wetted surface The portion of a vessel that is below the water line. Varies with type of vessel and with the weight and design of the vessel.

Whirling arm A mechanical device consisting of a fulcrum and an arm that extends. A testing sample is mounted at the tip of the arm. As the machine spins, readings are taken to determine aerodynamic characteristics of the model. The precursor to the wind tunnel.

Wingsuit A one-piece suit worn by skydivers to reduce friction.

Wing-warping An early system for lateral control of an aircraft in flight. The Wright's discovered it by watching Turkey vultures fly. First demonstrated the effect by twisting a long, empty box.

Yaw Movement of the nose from side to side. Rotation around the vertical axis.

Yaw axis The vertical axis through an aircraft that is perpendicular to the body of the wings with its origin at the center of gravity and directed towards the bottom of the craft. Also known as the vertical axis.Associations/Awards

Aeronautical Society of Great Britain Founded in 1866 to further "the art, science and engineering of aeronautics." Now the Royal Aeronautical Society. First and rounding members include Wenham.

John Fritz Gold Medal Highest award in the engineering profession. Established in 1902. Recipients include: Kelvin, Bell, Edison, Nobel, Sperry, and Taylor.

NACA National Advisory Committee for Aeronautics. March 13, 1915 to October 1, 1958. Preceded NASA.

NASA National Aeronautics and Space Administration. Established in 1958. Grew out of NACA.

National Academy of Sciences Founded in the United States in 1863 to "investigate, examine, experiment, and report upon any subject of science."

RINA The Royal Institution of Naval Architects. Founded in 1860 in London to advance the art and science of ship design. Formerly the Institution of Naval Architects.

SNAME Society of Naval Architects and Marine Engineers. Founded in the United States in 1893. Professional society that advances marine engineering.

The Copley Medal Awarded by the Royal Society of London for "outstanding achievements in research in any branch of science..." Recipients include: Robins, Franklin, Smeaton, Faraday, Agassiz, Darwin, Pasteur, Cayley, and Stokes.

The Franklin Institute Established in the United States in 1824. Promotes science education and development.

The Royal Society Founded in the UK in 1660 "to recognize, promote, and support excellence in science." Members include: Newton, Darwin, Einstein, and Hawking.

References

Birdman, "BIRDMAN." http://www.bird-man.com.

Wayne's Word. In *Blowing in the wind: seeds & fruits dispersed by wind*, ed. Wolffia. 2012. http://waynesword.palomar.edu/pifeb99.htm.

U.S. Centennial of Flight Commission, "Centennial of Flight." http://www.centennialofflight.gov.

Hermann-Oberth Raumfahrt-Museum. http://www.kiosek.com/oberth/.

Annual report of the board of regents of the smithsonian institution showing the operations, expenditures, and condition of the institution for the year ending June 30, 1902. 1903. Vol. Doc. No. 484, Part 1 of some aeronautical experiments.

The papers of William Froude M.A., Ll.D., F.R.S. 1810–1879. 1995. ed. Duckworth, R.N., Captain, A.D. London: Institution of Naval Architects. Print.

Pilotfriend. 2000. 2012. www.pilotfriend.com/.

Louis-Sebastien Lenormand. 2011. *Encyclopedia Britannica*: Encyclopedia Britannica. Print.

Interview with John Anderson, Curator of Aerodynamics, Smithsonian Institution, National Air and Space Museum. 2011. ed. Interview with John Anderson, Curator of Aerodynamics, Smithsonian Institution, National Air and Space Museum. Washington, DC. Print.

Bbc History. 2012. BBC. 2012. http://www.bbc.co.uk/history/.

Abbell, Sir Westcott. 1955. "William Froude, M.A., Ll.D., F.R.S. A Memoir." *The Papers of William Froude, M.A., Ll.D., F.R.S. 1810–1879*. Institution of Naval Architects. Ed. Duckworth, R.N., Captain, A.D., xi–xiv. London: The Institution of Naval Architects.

Abell, Sir Westcott. 1955. "A Memoir." In *The papers of William Froude, M.A., Ll.D., F.R.S. 1810–1879*. ed. Duckworth, R.N., Captain, A.D., xi–xiv. London: The Institution of Naval Architects. Print.

Adair, Robert Kemp. 1944. *The physics of baseball*. 2nd ed. New York: HarperPerennial. Print.

Administration, National Aeronautics and Space. 2010 "The Beginner's Guide to Aeronautics". ed. Benson, Tom. http://www.grc.nasa.gov/WWW/k-12/airplane/bga.html.

———. "Airplanes". 2010. ed. Shaw, Dr. Robert, J. http://www.grc.nasa.gov/WWW/k-12/UEET/StudentSite/airplanes.html#partsofplane.

———. "Invention of the Airplane (1897–1905)." 2010. *Power point presentations*: NASA: Glenn Research Center. Print.

———. "The Beginner's Guide to Rockets". 2010. ed. Benson, Tom. http://www.grc.nasa.gov/WWW/K-12/rocket/bgmr.html.

Alexander, David. 2002. *Nature's flyers : birds, insects, and the biomechanics of flight*. Baltimore: Johns Hopkins University Press. Print.

Allison, David K., Keppel, Ben G., Nowicke, C. Elizabeth. *D.W. Taylor*. United States Government Printing Office. Print.

Anderson, John David. 1997. *A history of aerodynamics and its impact on flying machines*. Cambridge aerospace series. Cambridge, NY: Cambridge University Press. Print.

————. 2002. *The airplane: a history of its technology*. Reston, VA: American Institute of Aeronautics and Astronautics. Print.

————. 2004. *Inventing flight: the Wright brothers & their predecessors*. Baltimore, MD: Johns Hopkins University Press. Print.

————. 2008. *Introduction to flight*. 6th ed. Boston, MA: McGraw-Hill. Print.

Armstrong, W.P. 1999. *Blowing in the wind: seeds & fruits dispersed by the wind*. Information and images of seeds dispersed by the wind.10/30/11, 2011.

Baals, Donald D., William R. Corliss. 1981. *Wind tunnels of Nasa*. Nasa Sp. Washington, DC: Scientific and Technical Information Branch for sale by the Supt. of Docs., U.S. G.P.O. Print.

Barnaby, Kenneth C. 1960. "1860-1864 the First Five Years - Iron Ships." *The Institution of Naval Architects, 1860–1960; an historical survey of the institutions transactions and activities over 100 years*, 7–47. London: Royal Institution of Naval Architects in association with Allen and Unwin. Print.

Barnaby, K.C. 1960. *The Institution of Naval Architects 1860–1960: an historical survey of the institution's transactions and activities over 100 years*. London: The Royal Institution of Naval Architects. Print.

Benford, Harry. 1991. *Naval architecture for non-naval architects*. Jersey City, NJ: Society of Naval Architects and Marine Engineers. Print.

Benson, Tom. 2012. "Nasa". 2009. *Glenn Research Center*. http://www.grc.nasa.gov/.

Besant, W. H., Ramsey, A.S. 1934. *A treatise on hydromechanics*. 2 vols. London: G. Bell. Print.

Brenkus, John. 2010. *The perfection point*. New York, NY: Harper.

Enfield, Harper. 2010. Publishers Group UK distributor. Print.

Brunel, Isambard. 1870. *The life of Isambard Kingdom Brunel, civil engineer*. London: Longmans, Gree, and Co. Print.

Butler, Don. "Adventures in airflow: Part Ii." *Cars & Parts*: 8. Print.

Calero, Julián Simón. 2007. *The genesis of fluid mechanics, 1640–1780*. Studies in history and philosophy of science. Dordrecht, The Netherlands: Springer. Print.

Carlisle, Rodney P. 1998. *Where the fleet begins : a history of the David Taylor Research Center, 1898–1998*. Washington: Naval Historical Center. For sale by the U.S. G.P.O. Print.

Çengel, Yunus A., Cimbala, John M. 2006. *Fluid mechanics : fundamentals and applications*. Mcgraw-Hill series in mechanical engineering. Boston, MA: McGraw-HillHigher Education. Print.

Chanute, Octave. 1894. Progress in flying machines. *The American Railroad Journal*. Print.

Clary, David A. 2003. *Rocket man: Robert H. Goddard and the birth of the space age*. 1st ed. New York: Hyperion. Print.

Concrete Society, Royal Institution of Naval Architects.1977. *Concrete afloat : proceedings of the conference on concrete ships and floating structures organized by the Concrete Society in Association with the Royal Institution of Naval Architects and Held in London on 3 and 4 March, 1977*. London: Telford. Print.

Congress, Library of. 1900. "Diaries and notebooks: 1900–1901, Wilbur Wright." In *Diaries and notebooks*. Ed. Congress, Library of. Washington, DC: Manuscript Division. Print.

————. 1900. "The Wilbur and Orville Wright papers." *Octave Chanute papers, 1902–10*. Manuscript Division. Print.

————. 1901. "The Wilbur and Orville Wright Papers." *Subject file: business journals and ledgers*. Manuscript Division. Print.

————. 1910. "The Wilbur and Orville Wright papers." *Scrapbooks, 1902–1914*. Manuscript Division. Print.

————. 1912. "The Wilbur and Orville Wright papers." *Wind tunnel—correspondence, 1912–1928, 1938–1946*. Washington, DC 20540: Library of Congress, Manuscript Division. Print.

————. 1920. "The Wilbur and Orville Wright papers." *Legal cases—montgomery v. Wright-Martin Ariway Corp.-depositions: Wright, Orville, 1920*. Manuscript Division. Print.

————. 1920. "The Wilbur and Orville Wright papers." *General correspondence: Taylor, David W., 1925*. Manuscript Division. Print.

Corner, E.D.S., E.J. Denton, and G.R. Forster. 1969. On the buoyancy of some deep-sea sharks. *Proceedings of the Royal Society of London Biological Sciences* 171(1025): 415–429.

Crouch, Tom D. 1989. *The Bishop's boys : a life of Wilbur and Orville Wright*. New York: W.W. Norton. Print.

Darrigol, Olivier. 2005. *Worlds of flow: a history of hydrodynamics from the Bernoullis to Prandtl*. Oxford, NY: Oxford University Press. Print.

Discovery Communications. 1999. "The Ultimate Guide: Dolphins". *The ultimate guide*. Discovery Communications, Inc. http://animal.discovery.com/features/dolphins/dolphins.html.

Earhart, Family of Amelia. 2012. "Amelia Earhart". http://www.ameliaearhart.com.

Eckert, Michael. 2006. *The dawn of fluid dynamics : a discipline between science and technology*. Weinheim: Wiley-VCH. Print.

Ferreiro, Larrie D. 2007. *Ships and science: the birth of naval architecture in the scientific revolution, 1600–1800*. Transformations: studies in the History of Science and Technology. Cambridge, MA: MIT Press. Print.

Fish, Frank. 2012. In Buoyancy in marine mammals and reptiles. ed. Hagler, Gina2012. Print.

Foundation, The National Science. "Science Nation". 2012.

Froude, William. 1861. "On the Rolling of Ships." In *The papers of William Froude, M.A., Ll.D., F.R.S. 1810–1879*. Institution of Naval Architects. ed. Duckworth, R.N., Captain, A.D., 40–64. London: The Institution of Naval Architects. Print.

———. 1862. "On the rolling of ships - appendices." N *The papers of William Froude, M.A., Ll.D., F.R.S. 1810–1879*. Institution of Naval Architects. ed. Duckworth, R.N., Captain, A.D., 65–76. London: The Institution of Naval Architects. Print.

———. 1868–1870."Observations and suggestions on the subject of determining by experiment the resistance of ships—correspondence with the admiralty." In *The papers of William Froude, M.A., Ll.D., F.R.S. 1810–1879*. Correspondence with the admiralty. ed. Duckworth, R.N., Captain, A.D., 120–128. London: The Institution of Naval Architects. Print.

———. 1869. "The state of existing knowledge on the stability, propulsion and sea-going qualities of ships, and as to the application which it may be desirable to make to her majesty's government on this subject." In *The papers of William Froude, M.A., Ll.D., F.R.S. 1810–1879*. Institution of Naval Architects. ed. Duckworth, R.N., Captain, A.D., 129–133. London: The Institution of Naval Architects. Print.

———. 1872. "Experiments on the surface-friction experienced by a plane moving through water." In *The papers of William Froude, M.A., Ll.D., F.R.S. 1810–1879*. Institution of Naval Architects. ed. Duckworth, R.N., Captain, A.D., 138–146. London: The Institution of Naval Architects. Print.

———. 1873. "Description of a machine for shaping models used in experiments on forms of ships." In *The papers of William Froude, M.A., Ll.D., F.R.S. 1810–1879*. Institution of Mechanical Engineers. ed. Duckworth, R.N., Captain, A.D., 206–212. London: The Institution of Naval Architects. Print.

———. 1874. "On Experiments with H.M.S. Greyhound." In *The papers of William Froude, M.A., Ll.D., F.R.S. 1810–1879*. Institution of Naval Architects. ed. Duckworth, R.N., Captain, A.D., 232–256. London: The Institution of Naval Architects. Print.

———. 1876. "The fundamental principles of the resistance of ships." In *The papers of William Froude, M.A., Ll.D., F.R.S. 1810–1879*. Royal Institution of Great Britain. ed. Duckworth, R.N., Captain, A.D., 293–310. London: The Institution of Naval Architects. Print.

———. "Experiments upon the effect produced on the wave-makiing resistance of ships by length of parallel middle body." In *The papers of William Froude, M.A., Ll.D., F.R.S. 1810–1879*. Institution of Naval Architects. ed. Duckworth, R.N., Captain, A.D., 311–319. London: The Institution of Naval Architects. Print.

———. "On the soaring of birds." In *The papers of William Froude, M.A., Ll.D., F.R.S. 1810–1879*. Royal Society of Edinburgh. ed. Duckworth, R.N., Captain, A.D., 340–344. London: The Institution of Naval Architects. Print.

Fuller, John. 2012. "How stuff works". 1998–2012. *Top 10 bungled attempts at one-person flight.* http://science.howstuffworks.com/transport/flight/classic/ten-bungled-flight-attempt3.htm.

Gale Group, Gale (Firm). 1999. *Gale biography in context.*

Garber, Stephen. 2011. "Centennial of Flight". *Born of dreams—inspired by freedom.* Various 2011/2012. http://www.centennialofflight.gov.

Garnerin, A. 1802. M. Garnerin's account of his ascent from St. George's parade, North Audley Street, and Descent with a Parachute, Sept 21, 1802. *The European Magazine and London Review,* July to December. Print.

Gawn, Andrew. 1955. "An evaluation of the Work of William Froude." In *The papers of William Froude, M.A., Ll.D., F.R.S. 1810–1879.* ed. Duckworth, R.N., Captain, A.D. London: The Institution of Naval Architects. Print.

Gibson, Andrew, Arthur Donovan. 2000. *The Abandoned Ocean: a history of United States Maritime Policy.* Studies in Maritime History. Columbia: University of South Carolina Press. Print.

Great Britain. Ship and Marine Technology Requirements Board, National Maritime Institute (Great Britain). 1980. *Commercial sail: proceedings of a symposium held at the Royal Institution of Naval Architects on 14 June 1979.* London: The Dept. Print.

Hays, Graeme C., Marshall, Greg J., Seminoff, Jeff A. 2006. Flipper beat frequency and amplitude changes in diving green turtles, Chelonia Mydas. *Marine Biology* 150: 1003–1009.

Henderson, Carrol L. 2008. *Birds in flight: the art and science of how birds fly.* Minneapolis, MN: Voyageur Press. Print.

History, Smitnsonian National Museum of Natural. 2010. "Ocean Portal". Smithsonian.

Homans, Isaac Smith. 1860. *Hunt's merchants' magazine and commercial review* 804. Print.

Hovgaard, William. 1941. "Biographical Memoir of David Watson Taylor, 1864–1940." *National Academy of Sciences of the United States of America Biographical Memoirs* XXII Seventh Memoir. Print.

Institute of Marine Engineers, Royal Institution of Naval Architects, and Society for Underwater Technology. 1975. *In-Water maintenance on ships: a joint conference held on Wednesday, 15 January 1975.* Trans I Mar E, 1975: Series B. London: Marine Media Management for I.M.E. Print.

Institution of Mechanical Engineers (Great Britain). Power Industries Division, Nihon Kikai Gakkai, and Royal Institution of Naval Architects. 1992. *Cavitation: international conference, 9–11 December 1992, Robinson College, Cambridge.* Proceedings of the institution of mechanical engineers. London: Published for IMechE by Mechanical Engineering Publications. Print.

Jakab, Peter L. 1990. *Visions of a flying machine: the Wright brothers and the process of invention.* Smithsonian History of Aviation Series. Washington: Smithsonian Institution Press. Print.

Johnson, Richard W. 1998. *The handbook of fluid dynamics.* 1 vol. Boca Raton, FL: CRC Press. Print.

Kosmodemyansky, A.A. 1956. *Konstantin Tsiolkovsky: his life and work.* Moscow: Foreign Languages Pub. House. Print.

Lamb, Sir Horace. 1895. *Hydrodynamics.* University Press. Print.

Leonardo, Bill Gates. 2001. *Il Codice Leicester, Codice Hammer.* Napoli: A.E.S. Art Books. Print.

Licht, S.C., M.S. Wibawa, F.S. Hover, and M.S. Triantafyllou. 2009. In-line motion causes high thrust and efficiency in flapping foils that use power downstroke. *The Journal of Experimental Biology* 213: 63–71.

Lilienthal, Otto. 1896. "Flying as a Sport." *The American Magazine,* January 3. Print.

Loewer, H. Peter. 2005. *Seeds: the definitive guide to growing, history, and lore.* Portland, OR: Timber Press. Print.

Proceedings of the Royal Society of London. June 17, 1869 to June 16, 1870. Taylor and Francis. Print.

Lytkin, Vladimir. "Konstantin Tsiolkovsky—the Pioneer of Space Travel." *Encyclopedia Astronautica.* Print. http://www.astronautix.com/.

Mahan, A.T. 1987. *The influence of sea power upon history, 1660–1783*. New York: Dover Publications. Print.

McLellan, Dennis. 2010. "Walter Frederick Morrison dies at 90; Father of the Frisbee." Obituary. *Los Angeles Times*, February 13, 2010. Print.

Merrifield, C.W. 1870. "The experiments recently proposed on the resistance of ships." *Transactions of the Institution of Naval Architects*. Vol. XI., 80–93. London: Institution of Naval Architects. Print.

Minamikawa, Shingo, Naito, Yasuhiko, Sato, Katsufumi, Matsuzawa, Yoshimasa, Bando, Takeharu, Sakamoto, Wataru. 2000. Maintenance of neutral buoyancy by depth selection in the loggerhead turtle *Caretta Caretta*. *The Journal of Experimental Biology* 203: 2967–2975. Print.

NASA. "Nasa". *Educator guide—rockets*. http://www.nasa.gov/pdf/280754main_Rockets.Guide. pdf.

Paipetis, S. A., Marco Ceccarelli. 2010. *The genius of Archimedes -- 23 centuries of influence on mathematics, science and engineering: proceedings of an international conference held at Syracuse, Italy, June 8–10, 2010*. History of Mechanism and Machine Science. Dordrecht, NY: Springer. Print.

Parnell, Gergory S., Patrick J. Driscoll, Dale L. Henderson. 2008. *Decision making in systems engineering and management*. Wiley Series in Systems Engineering and Management. Hoboken, NJ: Wiley-Interscience. Print.

Perrin, W.F., Bernd G. Würsig, Thewissen, J.G.M. 2009. *Encyclopedia of marine mammals*. 2nd ed. San Diego: Academic Press. Print.

Petrovitch, Vassill. "Buran-Energia". 2006–2012. http://www.buran-energia.com/blog.

Ramel, Gordon. "Fish anatomy: the swim bladder". *Earthlife Web*. http://www.earthlife.net/fish/ bladder.html.

Rankine, William John Macquorn. 1869–1870. On the mathematical theory of stream-lines, especially those with four foci and upwards [abstract]. *Proceedings of the Royal Society of London* 18: 207–209. Print.

Rawson, K.J., Tupper, E.C. 2001. *Basic ship theory*. 5th ed. Boston, MA: Butterworth-Heinemann. Print.

Resources, South Carolina Department of Natural. 2012. "South Carolina Wildlife". 2012. http:// www.scwildlife.com.

Rouse, Hunter, Simon Ince. 1957. *History of hydraulics*. Iowa City, IA: Iowa Institute of Hydraulic Research, State University of Iowa. Print.

Sergeeva, Galina, Timoshevenko, Elena. 2012. Russian Space Web. 2012. http://www.russianspaceweb.com.

Slade, Stuart. 1998. "Towing tank tests". http://www.navweaps.com/index_tech/tech-010.htm.

Stauth, David. 2005. Research shows hummingbird flight an evolutionary marvel. Corvallis, OR. http://oregonstate.edu/ua/ncs/archives/2005/jun/research-shows-hummingbird-flight-evolutionary-marvel.

Stimson, Richard. 2012. Wrightstories.Com. 2001. http://www.wrightstories.com/history.html.

Swartz, Sharon M., Iriarte-Diaz, Jose, Riskin, Daniel K. 2007. Wing structure and the aerodynamic basis of flight in bats. *American Institute of Aeronautics and Astronautics*: 10. Print.

Swift, Earl. 2007. The man with the original plan. *The Virginian-Pilot*, May 27, 2007. sec. Daily Break. Print.

Taylor, David Watson. 1900. *The United States Experimental Model Basin*. Society of Naval Architects and Marine Engineers. Print.

———. 1943. The principle of similitude. *The speed and power of ships; a manual of marine propulsion*. 3rd ed., ix, 301 p. Washington: U.S. Govt. Print. Off. Print.

Taylor, D.W., Sperry EA. 1909–1930. Taylor-Sperry correspondence. Print.

Taylor, David Watson, United States. Maritime Commission [from old catalog], and United States. David W. Taylor Model Basin Carderock Md. [from old catalog]. *The Speed and Power of Ships; a Manual of Marine Propulsion*. [3rd ed. Washington: U.S. Govt. Print. Off., 1943. Print].

Tokaty, G.A. 1994. *A history and philosophy of fluid Mechanics*. New York: Dover. Print.

Tupper, E.C. 2004. *Introduction to naval architecture*. 4th ed. Amsterdam, Boston: Elsevier, Butterworth Heinemann. Print.

Verne, Jules. 2007. *Journey to the Center of the Earth*. Print.

———. 2011. *Around the world in 80 days*. Kindle Edition. Print.

———. 2011. *From the earth to the moon*. Kindle Edition. Print.

Videler, John J. 2006. *Avian flight*. Oxford University Press. Print.

Wendt, John F., John David Anderson, Von Karman Institute for Fluid Dynamics. 2010. *Computational fluid dynamics: an introduction*. Berlin, NY: Springer-Verlag. Print.

Wright, Wilbur, et al. 2001. *The papers of Wilbur and Orville Wright: including the Chanute-Wright letters and other papers of Octave Chanute*. 2 vols. New York: McGraw-Hill. Print.

Wright, Wilbur, Orville. 1909. Story of our lives. *New York Herald,* March 5, 1909. Print.

Author Index

A
Ader, Clement, 99
Archimedes, 25, 45, 65, 67
  buoyancy, 67
Archytas, 85
  "The Dove," 85
Aristotle, 65–69, 71
  continuum, 66

B
Bacon, Roger, 101
Bernoulli, Daniel, 5, 6, 23, 26–28, 32, 41, 51,
    60, 75, 79, 89, 90, 173
  Bernoulli principle, 6, 22, 28, 51, 173
Blanchard, Jean-Pierre-Francois, 93
Boyle, Robert, 70–72
Brunel, Isambard Kingdom
  SS Great Britain, 77, 78, 109
  SS Great Eastern, 77, 79, 80, 109, 110, 112
  SS Great Western, 77, 78, 109, 110

C
Cayley, George
  camber, 166, 167
  Cayley's coin, 95
  drag, 163, 165, 166, 210
  gravity, 57, 166–168
  lift, 95, 97, 163, 165–171, 210
  thrust, 57, 166, 167, 210
  whirling arm, 161, 163, 166, 167
Chanute, Octave, 95, 98, 99, 162, 170, 171,
    176, 177, 182, 184, 187–190, 192–195

progress in flying machines, 95, 162,
    170, 184
Wright brothers, 95, 165, 171,
    184, 195
Congreve, William, 102

D
da Vinci, Leonardo
  AV=constant, 69, 70
  Codex Atlanticus, 86, 87, 160
  Codex Leicester, 69
  Codex on Flight, 86, 87
  ornithopter, 26, 86
  wind tunnel principle, 87, 170
de Fontana, Joanes, 101
de Rozier, Pilatre, 92, 93

E
Earhart, Amelia, 100
Euler, Leonhard
  Euler equations, 75, 77, 224

F
Froude, William
  birds, soaring, 11, 13
  Froude number, 80, 131
  Greyhound trials, 128, 129
  law of similitude, 80
  rolling of ships, 80, 110, 112,
    115, 131
  Russell, John Scott, 110, 111

G. Hagler, *Modeling Ships and Space Craft: The Science and Art of Mastering the Oceans and Sky*, DOI 10.1007/978-1-4614-4596-8,
© Springer Science+Business Media, LLC 2013

Subject Index

A

Aeolpile, 68
Aerodromes, 95, 96, 174–176, 192
Aspect ratio, 13, 14, 17, 28, 29, 162, 191, 192
AV=constant, 69, 70

B

Baseball, 32, 136
Bernoulli principle, 5, 6, 22, 26, 28, 32, 41, 51, 89, 173
Birds, soaring, 11
Blubber
    marine mammals, 46, 47
Boundary layer, 35, 39, 41, 55–59, 100, 101, 224–226
Buoyancy, 45–52, 67, 68, 92, 113, 114, 137, 216

C

Calculus of variations, 76
Camber, 6, 7, 26–28, 30, 36, 51, 89, 95–99, 163–167, 169, 170, 172, 174, 191, 192
Canada geese, 3, 4, 6, 11, 31, 60
Cayley's coin, 95
Codex Atlanticus, 86, 87, 160
Codex Leicester, 69
Codex on Flight, 86, 87
Computational fluid dynamics, 223–227
Continuum, 9, 55, 66, 76
Cyclists, 31, 32, 60

D

The Dove, 85
Drag, 5, 6, 9, 13, 15, 26–28, 31, 36–41, 54, 56–59, 90, 93, 100, 101, 123, 160, 162–166, 171, 182, 186, 190–192, 210, 213, 489

E

Euler equations, 224
Experimental model basin, 82, 130, 136, 137, 139–141, 143–146, 150

F

Flying insects, 18–19
Flying squirrels, 21, 30, 31
Frisbee, 35–37
From Earth to the Moon, 102, 208
Froude number, 80, 131

G

Gravity, 3, 5, 7, 15, 25–27, 30, 36, 45, 48, 50–57, 60, 69, 73, 88, 113–115, 125, 166–168, 173, 199, 200, 213, 216
Gray's Paradox, 56
Greyhound trials, 128, 129

H

Hummingbirds, 15–17

G. Hagler, *Modeling Ships and Space Craft: The Science and Art of Mastering the Oceans and Sky*, DOI 10.1007/978-1-4614-4596-8, © Springer Science+Business Media, LLC 2013